VHF Propagation

A Practical Guide For Radio Amateurs

By

Ken Neubeck, WB2AMU
Gordon West, WB6NOA

CQ Communications, Inc.

Library of Congress Control Number 2004102509
ISBN 0-943016-30-4

Editor: Edith Lennon, N2ZRW
Layout and Design: Elizabeth Ryan

Front Cover: Aurora over Chena Hot Springs, Alaska, February, 14, 2004. Shown in the foreground is the aptly-named Ice Hotel, constructed entirely out of ice. (Photo by Chip Margelli, K7JA)

Published by CQ Communications, Inc.
25 Newbridge Road
Hicksville, New York 11801 USA

Printed in the United States of America.

PREFACE

WHY DO WE NEED A BOOK ON VHF PROPAGATION?

Radio amateurs are very lucky to have some of the best prime radio frequency real estate in the VHF spectrum where major excitement can occur!

The VHF spectrum begins at 30 MHz and goes up to 300 MHz. For most of the world, this includes the amateur radio VHF bands of 6 meters (50 MHz), 2 meters (144 MHz), 220 MHz, and 432 MHz. In addition, there is the 4-meter band (70 MHz), which is used in some parts of Europe and South Africa. Just about all radio amateurs in the United States, as well as those in many other areas of the world, have access to these VHF radio frequencies, regardless of license class. And the propagation modes that are often seen on these frequencies make for some pretty incredible operating conditions. So learning about VHF propagation can add tremendously to your enjoyment of the hobby!

It is the authors' strong opinion that observing the kaleidoscope of different propagation modes found in the VHF range will lead to a clearer understanding of the modes in general, including when they appear on the HF and UHF amateur radio bands. When certain propagation modes appear on VHF, it is usually a singular event as compared to the multiple propagation paths and modes that appear on a busy HF band, such as 15 or 20 meters.

Most of the modes that appear on VHF also appear on HF, but not always at the same intensity or with the same effects. And, as opposed to the dependence of HF activity on the sunspot cycle, there are a number of propagation modes that appear primarily on VHF that are independent of the 11-year sunspot cycle. For example, propagation modes like sporadic-*E* (*E*s) and meteor scatter (Ms) are seen during certain times of the year without much influence from the solar cycle. So when F_2 propagation on 10 and 15 meters is severely impacted during the decline of the solar flux and sunspot count, there is still skip activity to bring excitement to VHF bands like 6 and 2 meters.

The two most common modes that hams will discover with regular monitoring on the VHF bands are tropo on 2 meters and *E*s on 6 meters. This book will cover both tropo and *E*s in sufficient detail that the VHFer will be able to recognize such events when they happen.

The combined ham radio experience of Ken Neubeck, WB2AMU, and Gordon West, WB6NOA, represents many years of VHF observations and research.

Ken has written extensively on many of the propagation modes that appear on 6 meters, and he has amassed significant data from his observations. His articles have appeared in *CQ* magazine, *CQ VHF*, *QST*, and *Worldradio*. He is the author of the book, *Six Meters, A Guide to the Magic Band* (Worldradio Books), which is now in its third edition.

Gordon is well known by many hams for the ham radio course he has constructed and taught throughout the country. His travels have taken him through many parts of the United States, where he has observed VHF phenomena firsthand. Gordon has written for all of the amateur radio magazines during the course of his career, including his current writings for *CQ* and *CQ VHF* magazines. Gordon, or "Gordo," is one of most recognizable ham radio personalities in our hobby today.

Ken is based in Long Island, New York, and Gordo is based in Southern California, below Los Angeles. Bringing these two different perspectives to the study of VHF propagation is also extremely helpful, as location plays a very important part in VHF operation.

It is the opinion of the authors (and many others) that much of the groundbreaking advances taking place in the world of radio are happening in the VHF and lower UHF ranges. For example, new transmission modes in the area of digital communications, such as JT-65 and high-speed CW (HSCW), were developed in response to the weak-signal modes that were seen on these frequencies.

Despite these groundbreaking advances, however, the VHF bands are not utilized to their maximum capabilities by the vast majority of radio amateurs. Too often, the VHF bands have been thought of as just repeater frequencies. As a result, there is a dearth of weak-signal activity (SSB and CW) on a daily basis on several of the bands, despite equipment availability. And the 220-MHz band has only a minimal amount of equipment available for SSB or CW work, which has impacted use of this band.

But, as you'll learn in these pages, VHF offers so much more than just repeater activity!

This book will reference the worldwide grid square system that was originally developed in Europe and used by VHFers as part of their contact exchange information. This is a standard convention, and complete maps are available from the ARRL or the Internet. The value of the grid square becomes evident when choosing a beam heading.

Some math, in the form of trigonometry and algebra, is used in the book, but this is kept to the level necessary for the reader to understand. The writing style is narrative to provide an enjoyable read.

Also available, as a companion to the written information you'll find here, is an audio course on cassette or CD. This will provide the reader with actual audio samples of how the different modes discussed actually sound. This will assist operators new to VHF propagation in identifying and working these modes. You'll find details on the course in Appendix 2.

So, there you have it, a book that is totally devoted to VHF propagation, written by experienced hams who love to operate on the VHF bands. There will be a balance of scientific data and practical observations that is not found in any other book regarding VHF propagation. It is the authors' hope that the information presented here will encourage the amateur radio population to make greater use of these fascinating bands!

Ken Neubeck, WB2AMU, and Gordon West, WB6NOA

ACKNOWLEDGEMENTS

The authors wish to acknowledge the tremendous support for this project from the CQ staff, especially Dick Ross, K2MGA, Rich Moseson, W2VU, Edith Lennon, N2ZRW, and Elizabeth Ryan. Their assistance helped make this book happen. We also wish to thank the many radio amateurs who are referenced throughout the book for their invaluable input. Also, our sincere thanks go to our wives, Fran Neubeck, and Suzy West, N6GLF.

We had a lot of fun working on this book and hope that you'll enjoy it.

MEET THE AUTHORS

Ken Neubeck, WB2AMU, is an active HF and VHF operator. He is a contributing editor for *CQ VHF* magazine and the author of many articles for *CQ*, *CQ VHF*, *QST*, and *Worldradio* magazines. In addition, he has written the definitive book on 50 MHz: *Six Meters, A Guide to the Magic Band* (Worldradio Books). Ken has also written books on military aircraft, such as the A-10 and F-105, as well as a textbook for reliability engineering that is a standard for the industry.

Gordon West, WB6NOA, is one of the best-known hams in the hobby today as a result of his classic body of work over the years, including his comprehensive amateur radio license exam study books, which continue to be a major source for newcomers to the hobby. Gordon is a contributing editor for both *CQ* and *CQ VHF* magazines and is a frequent speaker at amateur radio conventions and club meetings. He has also written extensively on topics relating to maritime VHF radio and ham radio for RV use, as well as on electronic topics for commercial electronic magazines.

Table of Contents

ACURA

1919 Torrance Blvd.
P.O. Box 2215
Torrance, CA 90509-9870

January 2006

Emissions Warranty Extension and Free Service Reminder

Dear Acura Owner:

We are sending you this notice to remind you that your vehicle is covered by an **emissions warranty extension** and to encourage you to take advantage of the **free** one-time 75,000-150,000 mile inspection and parts replacement offer detailed below. Because your Acura is about 9 years old, it (on average) should fall within that mileage range. If your vehicle still has less than 75,000 miles, and it has never had the **free** 50,000-75,000 mile emissions control system inspection (also detailed below), please have it done at your convenience.

Background

In 1998, Acura entered into an agreement with the U.S Environmental Protection Agency (EPA) and the California Air Resources Board (CARB). The reason for this agreement was an assertion by the EPA and CARB that there is a technical problem with your Acura's On-board Diagnostic system (OBD) that causes it to not comply with the Federal Clean Air Act and California regulations. OBD is a required part of your vehicle's computer system that monitors the functions of the emissions control system.

As part of this agreement, Acura is providing owners of certain 1995, 1996, and 1997 vehicles these benefits:

1. The Federal and California Emissions Warranties have been extended to 14 years from the date your vehicle was first sold, or 150,000 miles, whichever comes first.

2. When the vehicle has between 50,000 and 75,000 miles, an Acura dealer will inspect your vehicle's emissions control system free of charge. Any warranty repairs or replacements needed on your emissions control system will be done free of charge, including all parts and labor.

3. When the vehicle has between 75,000 and 150,000 miles, an Acura dealer will again inspect your vehicle's emissions control system and make any needed warranty repairs or replacements free of charge. In addition, the dealer will replace the spark plugs, ignition

What You Should Do

1. If your vehicle fails a state Smog Check, or the Malfunction Indicator Lamp comes on (the amber Check Engine light on the instrument panel) take your vehicle to an Acura dealer. Under the terms of the warranty extension, the dealer will repair any emissions-related problem free of charge (including diagnosis, parts, and labor). You should maintain your vehicle according to the maintenance schedule in your owner's manual. This schedule is designed to keep your vehicle's emissions control systems functioning properly.

2. If your vehicle has between 50,000 and 75,000 miles, contact your local Acura dealer to make an appointment for an emissions control system inspection. The inspection and any necessary repairs are free of charge.

3. If your vehicle has between 75,000 and 150,000 miles, contact your local Acura dealer to make an appointment for an emissions control inspection, the replacement of ignition parts noted above, and an oil and oil filter change. All inspections, repairs, and replacements are free of charge. We recommend that you have this work done in conjunction with the normal scheduled maintenance at either 90,000 or 105,000 miles. Depending on the model, either 90,000 or 105,000 miles is the normal interval for spark plug replacement. Please refer to the maintenance schedule in your owner's manual.

If you have paid for parts/repairs which you believe should have been eligible for coverage under this emissions warranty extension, please submit copies of the repair orders and any supporting documentation for consideration to:

American Honda Motor co., Inc.
Acura Client Services
P.O. Box 2964
Torrance, CA 90509-2964

Please note that only emissions-related parts/repairs may be eligible. Your warranty book contains a list of emissions related parts. If you have lost or misplaced your warranty book, a replacement copy can be downloaded from www.acura.com (click on OWNERSHIP EXPERIENCE, OWNER LINK, WARRANTY/CONTRACTS).

If you have any questions about this notice or your emissions warranties, please call Acura Client Services at (800) 382-2238.

Sincerely,

AMERICAN HONDA MOTOR CO., INC.
Acura Automobile Division

CHAPTER 1

INTRODUCTION TO THE TROPOSPHERE AND THE IONOSPHERE

Without the magic that goes on in the earth's ionosphere that allows for radio waves to be reflected, there would be no available means for radio signals to be propagated over significant distances. Imagine what our amateur radio hobby would be like if there were no propagation modes on the HF or VHF bands? Too often we take for granted that propagation exists and that radio waves are reflected off the ionosphere and returned to earth. If we did not have a formation of ions in the ionosphere for radio waves to be reflected off of, radio waves could be absorbed or could pass right through and go into space. We would simply be left with line-of-sight communications, which are limited to a fixed distance because of the curvature of the earth.

Regions of Propagation

To learn about how VHF propagation takes place, it is important to understand some of the different atmospheric layers and what effect they have on radio waves. The two major regions of the atmosphere that we will look at initially are the ionosphere and the troposphere.

We can see the general makeup of the earth's atmosphere in **Figure 1-1**. The chapters of this book have been arranged to begin with those propagation modes associated with the layers closest to the earth, beginning with line of sight and tropo, all the way up to the *E* and *F* layers of the ionosphere.

The troposphere is found up to approximately 20 kilometers above the earth. While most radio waves cannot be reflected in this region, the tropospheric area is capable of creating ducts for

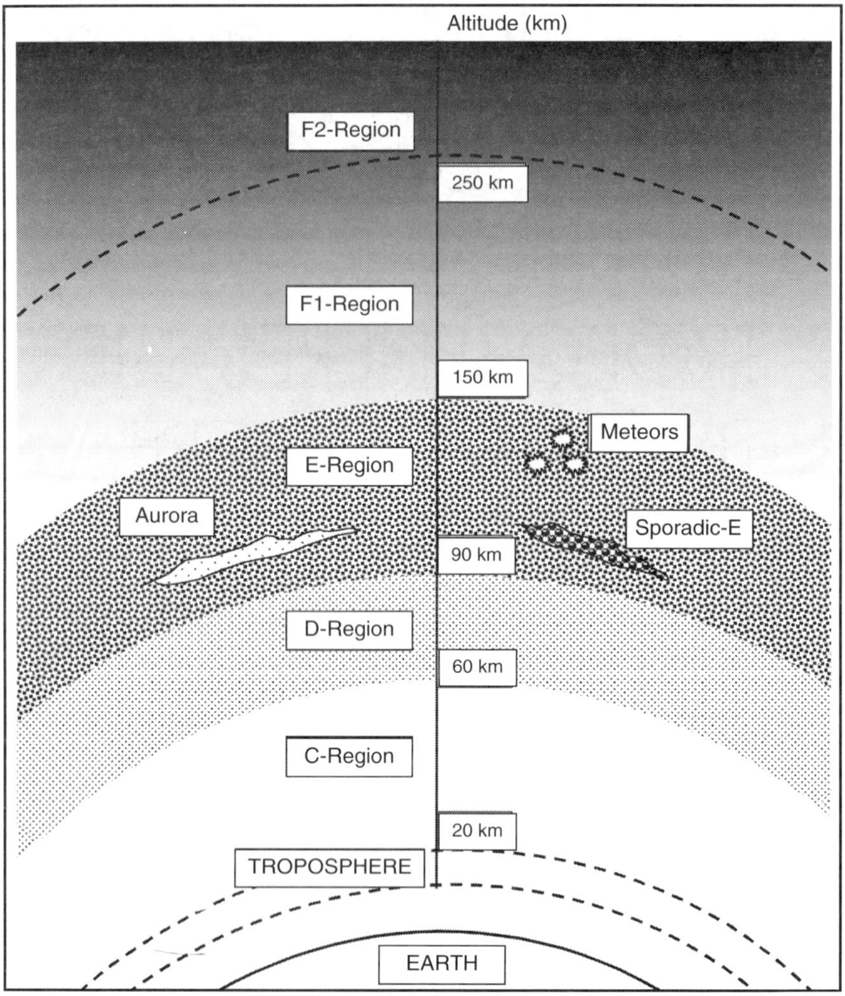

Figure 1-1. Regions of the ionosphere, also indicating major VHF modes of propagation.

certain frequencies of radio waves to travel through. This is the bread-and-butter mode of the VHF bands, and there are certain conditions that lead to enhancement of this ducting that will be covered in detail in **Chapter 2**.

Next, we'll examine the ionosphere itself. The lowest part of the ionosphere is the *D* region, which begins at 60 kilometers. This is an area that can either absorb radio waves or allow them to pass through, depending on their frequency. The *D* region is not known for either ducting or its reflective properties; however, its upper part does have an impact on some of the VHF modes of propagation in terms of absorption.

Things start to get interesting when we encounter the *E* region, which begins at about 90 kilometers. The *E* region is less dense than the *D* region, and it contains interesting ion particles that

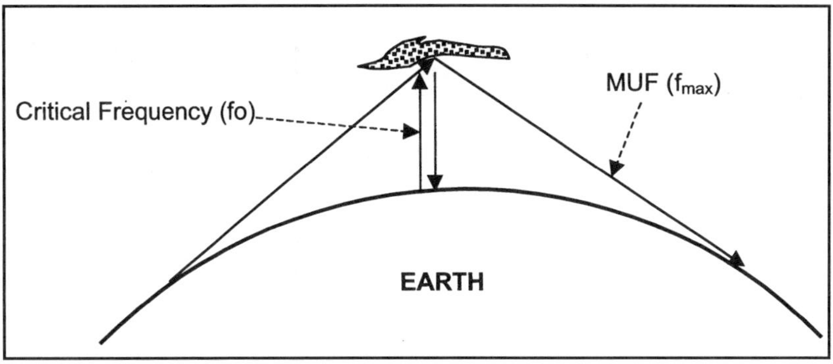

Figure 1-2. The relationship between critical frequency (f_0) and MUF (f_{max}).

are significant for VHF radio propagation. These ion particles are the key ingredients of the amazing sporadic-*E* (abbreviated *E*s) and mysterious aurora (abbreviated Au) phenomena. These are major VHF propagation modes on the 6-meter band and, on certain occasions, the 2-meter band. Another significant mode of propagation that occurs in the *E* region is meteor scatter (abbreviated Ms), which allows for short duration contacts to be made on several of the VHF bands. We will learn about all these modes in detail in **Chapters 3** through **5**.

Beginning at 150 kilometers, we enter into the lower half of the *F* layer. The area of interest to VHF operators is the F_2 layer, which starts at 250 kilometers. During the peak of the solar sunspot cycle F_2 propagation is a major mode on 6 meters and we will cover it in detail in **Chapter 6**. Another *F*-layer propagation mode we'll cover is transequatorial propagation (abbreviated TEP). This occurs over the geomagnetic equator and allows for some very long-range contacts to be made on 6 meters and, on rare occasions, 2 meters for certain areas of the world. We'll cover this mode in detail in **Chapter 7**.

Atmospheric Conditions

Part of understanding the ionosphere is recognizing that the makeup of these areas high above earth differs quite dramatically from region to region. For example, the *D* region (60 to 90 kilometers) experiences sub-freezing temperatures in the range of –70°C. Beyond 90 kilometers temperatures start to increase to about 2000°C at the 300-kilometer mark. In addition to the temperature variations, the molecular density of the *E* and *F* regions is low, compared to lower altitudes. So when ionization takes place in specific areas of these regions, the electron density increases, with the result that radio waves can be reflected.

Photo 1-A. This is the ionosonde station and associated antenna setup for NOAA at the Boulder, Colorado, location. The station sends a sounding signal into the ionosphere on an hourly basis and an on-site computer saves the resulting critical frequency value data that is collected. Ionosonde stations are typically housed in modest buildings in rural areas and are generally unmanned for much of the time. (Photo by Ken Neubeck, WB2AMU)

It is important to understand what conditions can cause radio waves to be reflected in the ionosphere. It may help to remember that radio signals generated by our radio transmitters are not a single line vector. Instead the waves have certain shapes depending on the style of antenna used. For example, the dipole antenna

Photo 1-B. Close-up view of the antenna setup for the ionosonde station in Boulder, Colorado. (Photo by Ken Neubeck, WB2AMU)

generates a donut pattern, with the antenna itself running through the hole of the donut. This donut pattern can reach for miles into the sky until it is either reflected or passes through into deep space. It also helps to visualize the ionosphere as a reflecting surface in the sky that can bounce radio waves back toward earth, depending on the density of the reflecting surface and the frequency of the radio wave. Normally, we tend to think of radio wave reflection in the conventional terms of straight lines and in simple geometric forms. However, while this visual aid is helpful, you should be aware that reflecting surfaces are

Band	Domestic Calling Frequency	Domestic Beacon Portion
50 MHz	50.125 (SSB and CW)	50.060–50.080
52 MHz	52.525 (FM)	-
144 MHz	144.200 (SSB and CW)	144.100–144.300
146 MHz	146.52 (FM)	-
222 MHz	222.100 (SSB and CW)	222.000–222.100
432 MHz	432.100 (SSB and CW)	432.000–432.100

Table 1-1. Important frequencies to monitor on VHF bands. Note: 50.110 is the DX calling frequency for 50 MHz.

not homogeneous and can be irregular in terms of surface and density. (This is particularly true with *E*-region propagation modes, such as *E*s and Au).

Some Tricks of the Trade

One important term often used when discussing radio wave propagation is Maximum Usable Frequency (abbreviated MUF, or f_{max}). This refers simply to the highest radio wave frequency that can be reflected off a formation in the ionosphere. You may have seen MUF prediction charts for HF propagation for a given time and month, which are typically plotted for MUFs reaching 28 MHz, or 10 meters. For VHF propagation, we're looking at MUFs that reach 50 MHz, or 6 meters, and higher.

In the observation of MUF, instruments are used to measure the vertical or critical frequency (abbreviated f_0) of a formation. The critical frequency is the highest frequency that can be reflected vertically and returned back to the source. One such measuring instrument is known as an ionosonde. An ionosonde typically transmits a two-minute sweep of frequencies up to 20 MHz at the top of each hour. The transmission is sent straight up into the ionosphere, and the frequencies of any reflections that are sent back are measured and recorded into a data file. There is a worldwide network of ionosonde stations, with one of the active U.S. stations located in Boulder, Colorado, under the supervision of NOAA (see **Photos 1-A** and **1-B**).

Based on the type of formation (or propagation mode), the MUF can be determined by the critical frequency observed. **Figure 1-2** shows the basic relationship between critical frequency and MUF. As we examine some of the VHF propagation modes, we'll show the unique mathematical relationship between MUF and critical frequency.

For many hams who may have been spoiled by the success of HF propagation, where it is relatively easy to make long-range contacts through ionospheric reflections, it may be initially somewhat disappointing when exploring VHF radio propagation (or lack thereof). Many of the HF bands are active on a daily basis, but the weak-signal areas of the VHF bands (located in the lower parts of each band) are less active, unless propagation is present. However, many hams stay the course, remaining on the VHF bands and accepting the challenge of patiently searching for VHF propagation.

The goal of this book is to make new and veteran hams alike more knowledgeable concerning the various VHF propagation modes and when they can occur so that they can take advantage of events as they happen. First you need to know where to start monitoring on each VHF band for activity; the calling frequency is usually a good place to start. **Table 1-1** shows the calling frequency as well as the beacon subband. Note that since the 6-meter band has international capabilities, there is a DX calling frequency (50.110 MHz) and an international beacon range.

CW beacons are invaluable aids for VHF propagation. There is a worldwide network of 6-meter beacons, located between 50.020 and 50.080 MHz, that's used by veteran operators. These beacons serve as an early warning system that a major propagation opening, such as *E*s or F_2, is occurring—often before actual human signals get on the band. Two meters has a beacon network as well and can alert operators to potential tropo openings. Beacons tend to be low power (less than 10 watts) with simple antennas, but they have proven their value over and over again.

What about equipment and modes of operation? The good news is that even modest equipment is suitable for taking advantage of certain propagation modes. For example, you could use a 5-watt FM handheld transceiver and work other repeater stations on 2 meters coming in via tropospheric ducting. QRP work (defined as 10 watts output or less for VHF) in the weak-signal portions of the band is also possible, even with simple antennas such as verticals and dipoles! Consistent success, however, on the VHF bands is usually achieved with moderate power levels of 100 watts and directional antennas. Remember, because of the smaller wavelength size associated with VHF bands, directional antennas, such as multi-element Yagis, are not that difficult to construct or put up. The use of directional antennas becomes more important with certain modes, such as Au, where hams in North America will be pointing their antennas north for maximum results.

Propagation modes are generally not discriminatory with regard to specific method of operation. For example, one can work *E*s propagation using the CW, SSB, FM, and digital modes. However, because of the fading associated with many of these modes of propagation, CW and SSB appear the most reliable. CW, in particular, is very useful when propagation is not that strong and good for propagation modes where distortion, such as aurora, is present. Knowledge of the code is, therefore, extremely handy at times for VHF work.

VHF propagation is also interesting in that there is significant spacing between the amateur radio bands in this range. While spacing between the different HF bands is on the order of 3 or 4 MHz (the HF bands being 1.8, 3.5, 5, 7, 10, 14, 18, 21, 24, and 28 MHz), the VHF bands are much wider apart! Specifically, there is 22 MHz separating the last HF band of 28 MHz and 50 MHz, the 6-meter band. There is a 90-MHz gap between 50 MHz and 144 MHz, and another 80 MHz of space before we reach 222 MHz. It is another 200 MHz from that band to 432 MHz.

With this amount of spacing between bands, it's easy to see why we end up describing some very different bands in the VHF spectrum. The higher VHF bands are better for tropo than is 6 meters, but six has better ionospheric propagation.

Now let's get started learning more about each of the individual VHF propagation modes and how they affect each band!

CHAPTER 2

BEYOND LINE OF SIGHT: TROPO

How far can you see on a clear day? This question can be rephrased for hams, with VHF propagation in mind, as "How far can I talk FM simplex, or how far can I work weak-signal CW or SSB?"

Ask any sailor how far he can see to the horizon or to a distant lighthouse, and he may pull out his *Bowditch Practical Navigator*, the classic book of maritime navigation. From this he can closely determine line-of-sight range based on his height above seawater and the height of the distant lighthouse, dutifully listed on the local charts. It is not a coincidence that visual range and VHF/UHF range calculations are almost identical. For VHF radio waves, we might calculate radio range by using a nomograph, a plot of curves with a straight line of interest drawn across it to read values (in this case radio ranges for different VHF frequencies).

Microwaves *normally* travel 1.1 times farther than the optical horizon. The formula used to calculate the distance of the radio horizon, in kilometers, is

$$D = \sqrt{17h}$$

where:

D = distance to radio horizon (in km)
h = height of antenna above water

As an example, for a space wave transmission, the radio horizon distance of a transmitting antenna with a height of 100 meters is calculated as

$$D = \sqrt{17 \times 100} = \sqrt{1700} = 41.23 \text{ km}$$

At the other end of the radio circuit, there may be a target of 64 meters in height, and the calculation would be

$$D = \sqrt{17x64} = \sqrt{1088} = 32.98 \text{ km}$$

If we add 32.98 kilometers to 41.23 kilometers, the minimum microwave range would be 41.23 + 32.98 = approximately 74 kilometers. This would be under “normal” conditions.

But what is “normal” for the visual, as well as the VHF/UHF, horizon? We often hear about the microwave radio horizon being 4/3 of the visual horizon. We can then conclude that VHF and UHF radio waves, on up to radar microwaves at super-high frequency, will “normally” exceed what we could actually see on a clear day staring out to sea.

The closest portion of our atmosphere, sometimes called the weather region, exhibits a decrease in air pressure with altitude in an approximately logarithmic manner: the higher we go, the less air pressure there is. Air temperature also decreases with altitude, approximately 20 degrees Fahrenheit for every mile of increasing height, up to 40,000 feet. The number of water molecules also decreases with altitude, resulting in atmospheric density decreasing with height above the surface of the earth.

Light and radio waves are similar in properties because they are both electromagnetic waves, just at very different wavelengths. As both wave types travel out in a straight line, the top portion of the wave front in thinner air accelerates faster than the lower portion of the wave front in denser air. For light waves, the much shorter wavelength does not have the degree of bending that radio waves have at VHF and UHF frequencies. This is the reason for the 4/3 radio horizon over the visual horizon.

This bending of both light and radio waves is called refraction. The refraction of “normal” air is slightly higher than unity, around 1.000345 to1.000300. To determine refraction, or N, we use the following equation:

$$N = 10^{6(n-1)}$$

where:

N = refractive index value

A nomograph, indicating pressure, temperature, dew point, and the saturation of water vapor pressure in millibars will show how far VHF and UHF radio signals may travel.

Visual Cues

Those same factors (temperature, humidity, etc.) also contribute to some of the optical illusions you might see close to the horizon during periods of strange weather patterns. Increasing your knowledge of how these optical illusions occur will help you better understand why VHF, UHF, and microwaves take some strange paths over thousands of miles.

The most common optical illusion that may indicate expanded VHF/UHF range is super refraction. This occurs with certain types of weather conditions involving cold and hot air masses, and results in objects well over the horizon mysteriously popping up into view. No, the earth has not become flat, but rather images well beyond the visual horizon are refracted (or bent) within view, much like VHF, UHF, and microwaves refracting well beyond horizon line-of-sight calculations.

One range extender of very short radio and light wavelengths is simple scattering. Clap two blackboard erasers together, and then shine a flashlight through the chalk dust, and you have a great example of optical scattering. Up at the X-band (10,000 MHz), pairs of ham radio operators have completed otherwise impossible paths by each aiming their dish antennas at one enormous thunderhead and making a successful scatter contact.

Another common radio range extender may easily be seen as a ground mirage (from the French mirer, "to look at") and indeed it acts much like an optical mirror. This common mirage on hot, sunny days out in the desert makes the pavement ahead look like shimmering blue water. Actually, the air just above the pavement is super-heated into a thin band, and this band exhibits an extremely sharp change in refractive index (the measurement of refraction for different media), bending the blue sky above, mirror-like, and making it appear as blue shimmering water down at pavement level. But did you ever notice this only occurs on hot, almost-windless days? These are the same types of days and weather conditions that will bend our regularly line-of-sight radio waves over hundreds or even thousands of miles.

While the exciting longer-than-usual radio range on VHF we might experience under these conditions is fantastic, actually *seeing* these anomalies of refraction within sharp boundaries of cool and warm air can sometimes be downright shocking!

Imagine the reaction of our troops in World War II as they headed home towards Oahu, Hawaii, finally approaching Pearl Harbor. One moment they're looking at Diamond Head and palm trees, when a mile closer Oahu appears to break in two, with nothing but water in between! The city of Honolulu sinks under the horizon, where 20 minutes ago it was within clear view. After 30 minutes, the two seemingly disconnected landmasses now appear to come together, and out of the blue sea emerges all the hotels at Waikiki. A sea-level mirage simply swallowed up the visible land with an overlying mirage capping what you would expect to see.

Why should mirages be of interest to VHF operators? Well, when they appear it means that VHF radio conditions will be enhanced. They are visual phenomena with a strong radio connection. The three common types of mirages of interest to radio operators are:

• Inferior Image—an inverted image below the erect image
• Superior Image—an inverted image above the erect image
• Fata Morgana—a complex combination of distorted erect and inverted images

Any of these three atmospheric anomalies will indicate changed radio conditions, even if radio operators are not able to tell the difference between them.

The inferior mirage occurs when cool air overlies warm air, and the illusion appears below the object's true position as if coming up from the horizon. Many radio amateurs consider this an indication of ducting created between the two temperature masses, much like a wave-guide where signals are channeled in a single direction. Visually, this may appear as a distant boat (normally too far to be seen over the horizon) popping up within view, with an upside down image of this same boat also in view. Hams in Key West, Florida, have reported seeing the buildings of Miami suspended on the horizon, and a mirror upside-down tunnel view of these same buildings shimmering above. There were also reports that VHF and UHF repeaters from Miami were also coming in just as loud and clear as the mirage seen on the horizon.

An instance of a superior mirage occurred off the coast of Southern California when Santa Catalina Island came into view, with an erect image and an inverted image below it, to residents of Santa Barbara. The island is well beyond line of sight from Santa Barbara, whose residents had never before seen it. They also reported that the VHF repeater on the mountaintop was so strong that it could be worked with nothing but a paper clip stuck into an HT antenna port! Chuck Northcut, W7SRZ, a retired U.S. Coast Guard officer and avid VHF/UHF operator, also reported hundreds of mirages encountered all the way up to Alaska when patrolling hundreds of miles off shore, well beyond line of sight. His Coast Guard pals have reported regular VHF and UHF television and FM stereo radio reception when stationed at Ocean Station November, halfway between Hawaii and California.

Fata morgana, the third type of mirage, is encountered on VHF/UHF and microwaves and is common in the northern latitudes. "This phenomenon is based on rare atmospheric conditions of calm air, when multiple alternating layers of calm, warm, and cool air collect over the waters and, when seen with

Photo 2-A. The ATV signal from KH6HME in Hawaii received in Southern California on 426.250 MHz via tropo ducting. (Photo by Gordon West, WB6NOA)

the sun behind and at a critical elevation of about 45 degrees, seaside villages and their environs along the Sicilian shore were vertically magnified and distorted as fantastic, awe-inspiring images in subtle coloring, definition, and shape," reported noted marine weather writer Michael Mooney[1]. "In Muir Glacier in Alaska, it presented a scene that could be interpreted to show buildings, towers, and masts of the city of Bristol, England, 2,500 miles distant."

Twenty-five hundred miles is also the distance between California and Hawaii where hams every July and August regularly make SSB, FM, and ATV contacts with each other. When Paul Lieb, KH6HME, tropo expert and holder of several VHF/UHF distance records, is asked whether he has ever seen Southern California hanging on the horizon from his 8,200-foot

Photo 2-B. Paul Lieb, KH6HME, at his famous VHF beacon station in Hawaii. (Photo by Sally Davis, KK7IO)

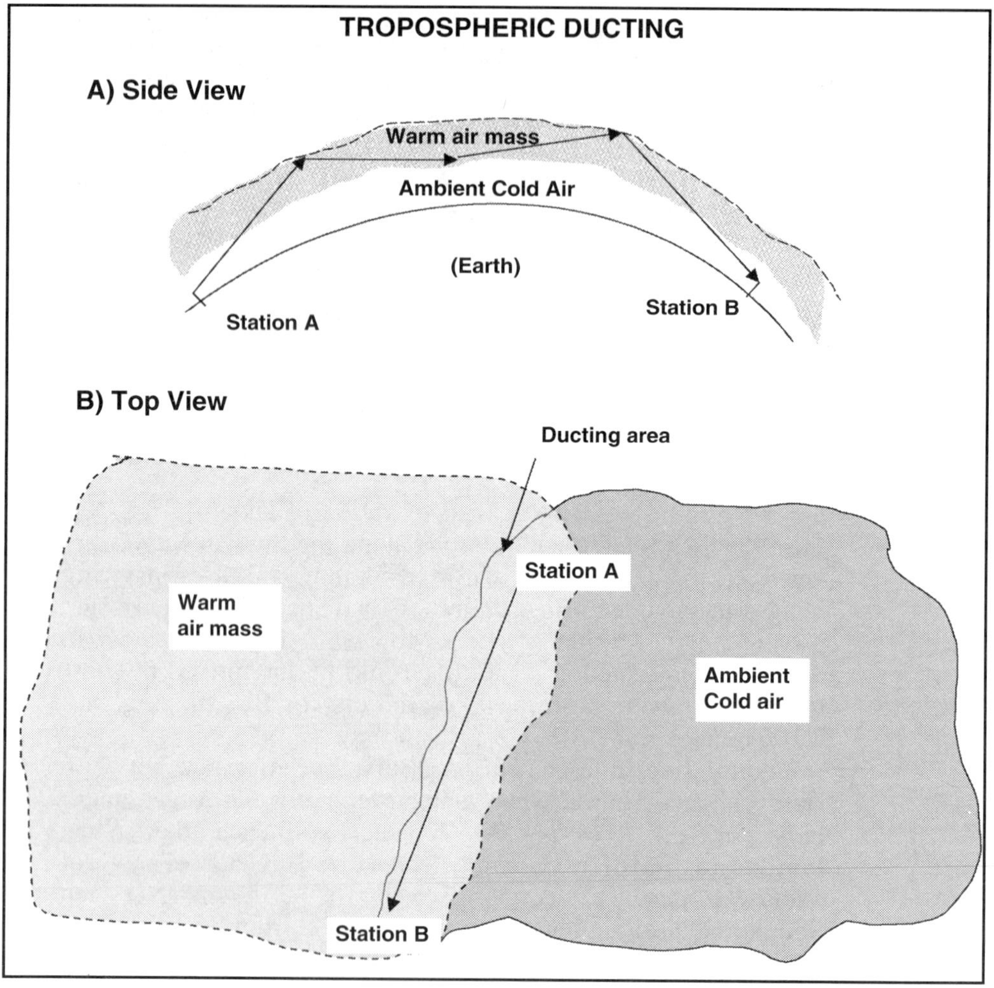

Figure 2-1. A typical tropo path can occur on the VHF bands through the formation of a duct between a warm air mass and a cool air mass. The warm air mass rides above the cold air mass during the collision of the two fronts, creating tropo ducting. This allows long-range contacts to be made between two stations as shown.

volcano view looking east, he smiles and tells other hams to come on up to the side of the volcano and see and hear for themselves (see **Photos 2-A** and **2-B**).

Ham radio operators define the conditions that lead to extraordinary long-range VHF/UHF and microwave contacts as tropospheric ducting attributed to high-pressure system subsidence and atmospheric stratification. Long-range VHF/UHF and microwave contacts are also possible along the line of slow-moving cold fronts approaching from the northwest or southwest and colliding with stationary warm air. **Figure 2-1** is a graphical representation of this tropo-ducting phenomenon.

Naval research laboratory Pacific Coast studies confirmed in the early 1960s that long-range VHF, UHF, and microwave propagation was indeed possible along ducts formed by trade wind inversion layers. One study, known as Project Tradewinds III, found that the 2,260-nautical mile propagation path between San Diego, California, and Opara, Hawaii, was weather related, with peak occurrences in July and August. Coincidentally, military frequencies chosen for the study were 220 MHz and 445 MHz, familiar ham bands to us today. The military concluded that the VHF/UHF and microwave radio duct is formed every summertime by the phenomenon of the general descent of high-altitude air into regions where it meets low-level sea air flowing in from the equator.

High-Pressure Tropo

Surface winds pick up moisture from the relatively cooler ocean, so there is a band of warm, moist air capped by stratified warmer dry air subsiding from the overlying static, high-pressure system. This cell of high pressure with associated subsidence is the key element for extraordinary long-range VHF/UHF and microwave tropospheric ducting. As the concentrated air from the high-pressure system begins to drop, it may become stratified approximately 1,000 feet above the ocean or land. When air becomes compressed, it begins to heat. The subsided air also loses moisture content, and with the little wind that is normally associated with a high-pressure system, a temperature inversion begins to form.

We call it an inversion because normally air gets cooler in higher altitudes, but becomes "inverted" in this stratified region associated with high pressure. Any time there is a sharp change in temperature, water content, and pressure, there may also be a steep change in the normal refractive index of air. This change in the refractive index may duct VHF and UHF signals for hundreds, and sometimes thousands, of miles within and just above the stratified layer. This ducting condition, unlike sporadic-*E*, has absolutely nothing to do with the ionosphere, and the condition may last for days and sometimes up to a week.

A temperature inversion, or even just an abnormally small decrease in temperature with height, will contribute an abnormally rapid decrease in N with height and contribute toward super refractivity. **Figures 2-2** and **2-3** present graphical examples of temperature and water inversion. According to Bruce Eggers, WA9NEW, an avid VHF propagation follower, what is required for a tropospheric duct is that a sufficiently super-refractive layer of sufficient depth lie above a layer of air, again of sufficient depth, with something of a near normal, or even

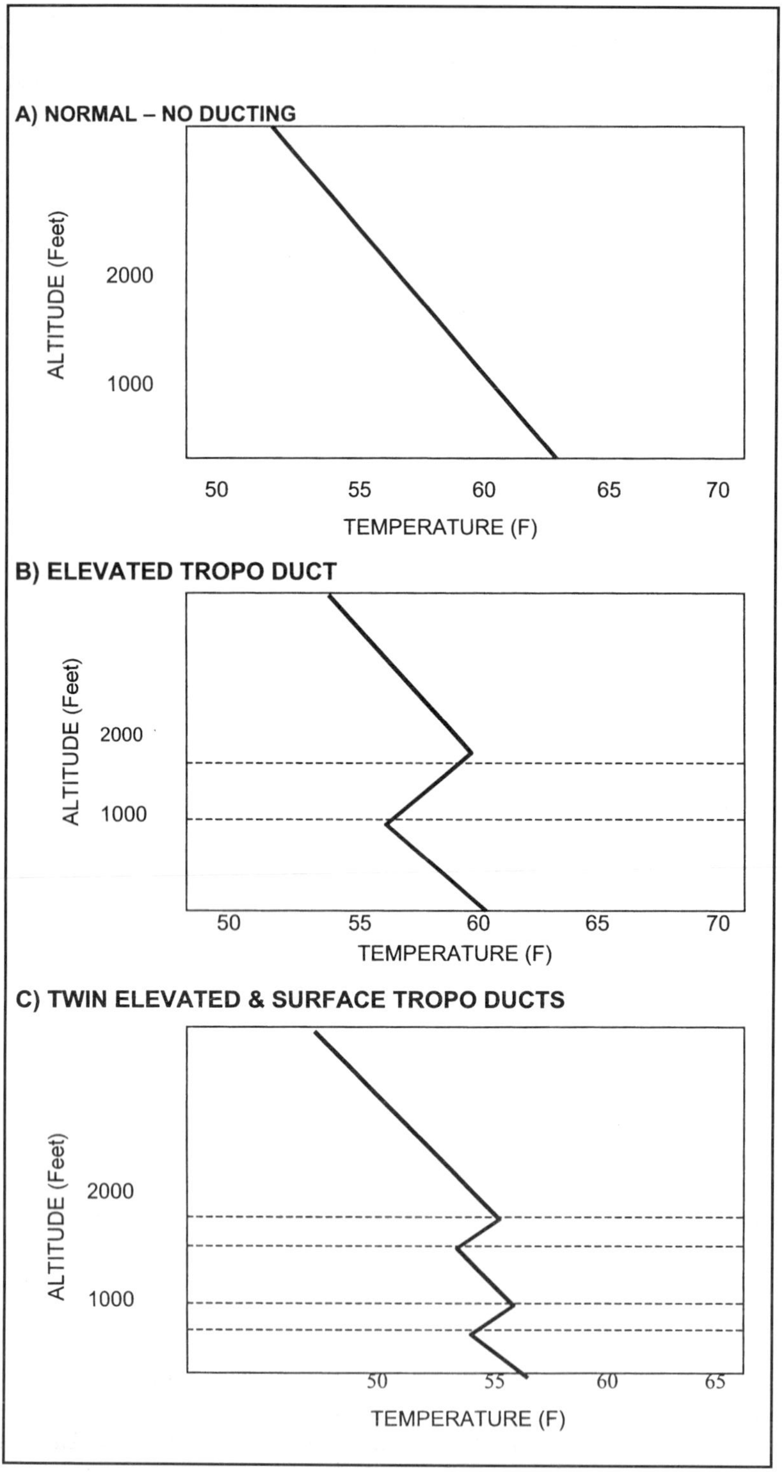

Figure 2-2. Types of temperature inversion.

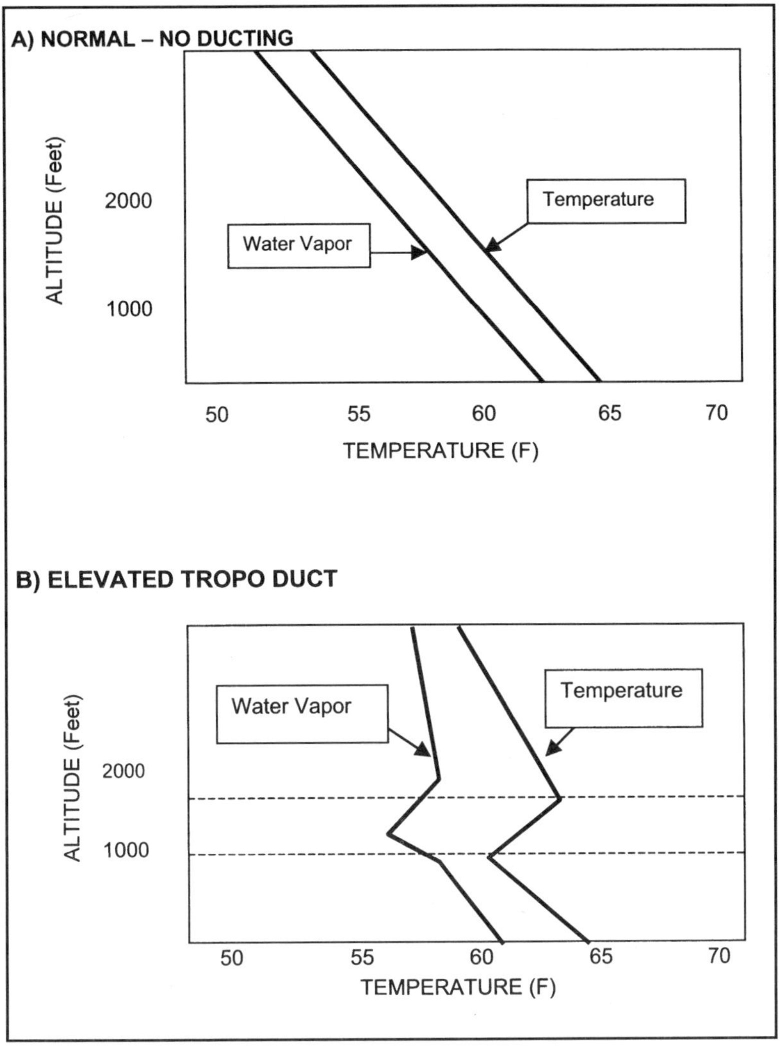

Figure 2-3. Types of temperature and water vapor inversion.

sub-refractive gradient of N. The formation of this duct is the result of the physics of the fluid. Any combination of parameters which contribute to a sufficiently thick layer of super-refractivity causes ducting. Under the right conditions, a transmitter 5,000 miles away has the same field strength as an identical transmitter only 70.7 miles away. There are a variety of different conditions which can produce ducting, including commonly occurring frontal layers and early morning radiational cooling of the lower layer of the atmosphere.

The depth of the inversion layer is very important. Layers which will not support observable ducting at 2 meters may well support ducting at higher frequencies (likewise, ducting may be experienced on 2 meters, but not on six). This is the situation that has been observed many times when the 2,500-mile Hawaiian beacon is barely heard on 2 meters in Southern California, yet the 432- and 1296-MHz beacons are extremely strong over the 2,500-mile path.

Tropospheric ducting experts point out many places on earth where high-pressure cells may contribute to long-distance tropospheric ducting. These include,

- The Pacific high between the West Coast of the U.S. and Hawaii
- The Atlantic high between the East Coast and Europe (which has yet to be worked on VHF and UHF)
- Summertime highs traveling west to east across the mid United States
- South Pacific Island highs, Indian Ocean high
- The regular high between South America and Africa

Here in the United States, July and August are the very best months for high-pressure system tropospheric ducting from subsidence and the stratified inversion layer. Your local weather forecaster will show you with weather maps the regional high-pressure systems that may have stalled. The weatherman may indicate unseasonably hot weather ahead with poor air quality. These are exactly the conditions that will give us terrific airwave quality from 50 MHz on up.

When you go outside and you see that band of brown air just hanging on the horizon, that is your visual cue that good tropospheric ducting conditions may exist. Depending on how far that band of brown air extends, VHF and UHF band conditions in the area could very well extend 100 to 500 miles.

Again, the inversion layer temperature *change* is a big contributor to what the VHF and UHF long-range band conditions will be like. For instance, a temperature increase of 20 degrees Fahrenheit per 1,000 feet of elevation could make the earth look almost flat. This is known as the "Hillingar Effect." For the Southern California-to-Hawaii duct to really get cooking on all bands, an abrupt change of 10 to 15 degrees within the stratified layer is sufficient to trigger good band conditions if this temperature difference remains undisturbed throughout the 2,500-mile path.

On rare occasions, along coastal boating regions, a 30-degree inversion up approximately 1,000 feet in elevation causes light rays to bend up in a bowl shape. This is a relatively localized condition that makes sailors think there's nothing but higher water all around them and that they're at the very bottom of the bowl. Waldemar Lehn, a professor of Electrical Engineering at the University of Manitoba, describes the phenomenon as follows: "The appearance of the foreground remains normal, but the horizon becomes elevated, and walls or barriers appear at the distance. It is simply an exaggerated case of the saucer-shaped earth where the apparent upward slope of the sea near the horizon is foreshortened, and the actual horizon appears as a wall only a few kilometers distant." VHF and microwave experts claim that this aberration to our local atmosphere will sometimes introduce as much as 6- to 9-dB path loss to an otherwise "slam dunk" point-to-point X-band system!

So there is little argument that local observed weather conditions may quickly tell the experienced tropo DX operator whether or not a long-distance path exists from 50 MHz on up.

William Alber, WA6CAX, a reserve aero squadron officer with his local county sheriff's department says that when the West Coast is open to Hawaii on 144 MHz and up he can actually fly right inside the inversion layer and watch a 10-degree climb in temperature, plus get a walloping signal from Hawaii. "I can smell it, I can see it, and I just about can guarantee it for a long-haul tropo opening," said Alber, reminding us that the visual atmospheric phenomenon predicts a sharp increase in received signals over long paths.

The tropo duct that forms between New York and the southern coastal states in August and September also has certain distinct visual characteristics. Frank, AA2DR, whose QTH is on Long Island, New York, told us that when he looks out and sees smoke hanging on the horizon in the direction of New York City, he can point his beam south and easily work all the way down the East Coast. Frank is well known for some extraordinary 2-meter and 432-MHz tropo contacts, and is also locally famous for his collection of VHF/UHF weak-signal radios and antennas. He gets local FMers turned on to the fascinating world of long-haul tropo ducting by conducting FM simplex activity nets on the 2-meter band (for more information, see the website www.liars.org).

It doesn't take complex daily meteorological forecasts to predict tropospheric ducting anywhere in the world. Many hams simply look at the regular weather forecast and "get a feel" for the local weather condition. Hams who are expert in tropo ducting are usually part of a local VHF/UHF weak-signal society, and the weekly nets usually bring up old-time tropo operators who can tell you what paths are workable and what paths just don't have a chance.

For instance, there is no known tropo duct between Los Angeles and Denver because of the mountains. Similarly, no tropo duct has ever been worked from Fairbanks, Alaska, down

Photo 2-C. Tropo is one of the few VHF propagation modes that has some visual indications in the atmosphere. This photo, taken some years ago from the top of the World Trade Center, shows a potential tropo duct developing in the haze over the northern part of New York City. (Photo by Drew Scelba, N2RFA)

to Seattle. However, these negative tropo propagation paths are in the minority. Following is a list of the summer and fall *regular* North America subsidence inversion paths (**Photos 2-C, 2-D,** and **2-E** show the physical appearance of some of these tropo paths):

- Hawaii to West Coast
- San Francisco to Los Angeles
- Los Angeles to Baja, California
- Texas to Florida
- Chicago to Texas
- Great Lakes to East Coast
- Northeast to Florida
- Florida to Bahamas
- Florida to Cuba
- South Carolina to St. Vincent and the Grenadines and South America

Again, hams are waiting for the day that a two-way tropo contact is made between the East Coast and Europe. The Bermuda-Azores high looks like a promising path for this to happen, according to Emil Pocock, W3EP, in an informative article on tropospheric ducting in the March 1996 issue of *QST* magazine. This particular high may ultimately be the key to the first transatlantic 2-meter QSO.

Photo 2-D. A view looking out to sea in Mexico. The haze over the water indicates conditions similar to those captured in Photo 2-C. (Photo by Gordon West, WB6NOA)

The most common number for local air pressure reported by both ends of a subsidence inversion tropospheric ducting contact is 1,026 millibars. But this subsidence inversion path will lead to some bizarre effects as subtle wind currents mix in with the stratified duct. VHF operators know the value of beacons that continuously transmit CW signals, and they are among the best ways of monitoring an incoming ducted DX signal. VHF operator Jim Ford, N6JF, has compared such a duct to a garden hose that is swishing back and forth in your driveway. You might have

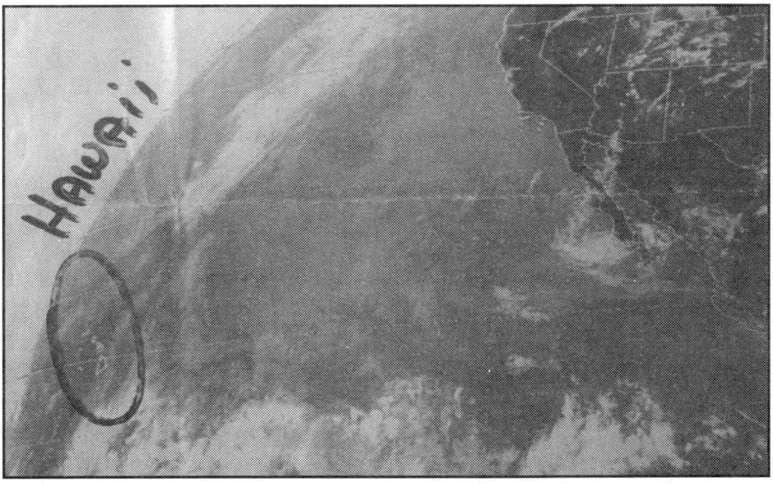

Photo 2-E. This weather fax plot shows the basic tropo path between Hawaii and California. Note the cloud formation above the path where the duct forms. (Photo of fax by Gordon West, WB6NOA)

Photo 2-F. This is the VHF beacon setup of Paul Lieb, KH6HME, on the big island of Hawaii. Paul has beacons running on several VHF bands (50.061, 144.170, and 432.075 MHz). These beacons are often heard during extended tropo path occurrences between Hawaii and California. (Photo by Gordon West, WB6NOA)

two stations separated by only 50 miles receiving an incoming tropo signal from 2,500 miles away. For one or two hours, both stations may hear the incoming signal relatively well, and then five minutes later, the signal gradually disappears from one station while gaining phenomenal signal strength at the other. Then both stations could lose the incoming signal. Yet a third station high up on a local mountaintop might begin to pick up the distant beacon (see **Photo 2-F** for an example of a beacon setup).

And you don't necessarily need height to be within the tropo duct. Several years ago during a California-to-Hawaii opening in July, almost no one could hear the beacon well until they drove down to the seashore and got within 10 feet elevation to the ocean surface. The duct was so strong that portable stations using the Yaesu FT-817 were working Hawaii at 2 watts of power output on 144 and 432 MHz into an omnidirectional loop antenna. All the while, stations that drove for hours to get high up on a mountaintop couldn't hear anything but the locals. And guess what? The next day, the mountaintoppers had more tropo than they could handle from many stations in Hawaii pouring into the West Coast, and those of us along the beach could barely hear a thing! (See **Photos 2-G** and **2-H** for examples of some low-power/simple antenna setups that can be used for strong tropo operations.)

Unlike sporadic-*E*, tropo signals have extremely gradual builds and fades, and it can take about an hour for a signal to build from just above the noise level to S9. And when the signal begins to change path, the decay gives you plenty of time to sign off with the distant station you're working and tell them to start looking for other stations in a slightly different part of the country. Sporadic-*E* can build and fade multiple times in a minute, but tropo slowly builds up, stays strong for hours and sometimes days on end, and then slowly fades out.

VHF Propagation Audio Course available. See Appendix II for details.

Another interesting thing about tropo is that there is a relationship between the size of the duct and what frequencies are

Photo 2-G. Making long-range contacts via tropo on the VHF bands is a popular activity in Southern California. Gordon, WB6NOA, and Chip, K7JA, are working stations on 432 MHz in the San Francisco Bay-area, about 400 miles away from them. (Photo by Gordon West, WB6NOA)

Photo 2-H. Suzy West, N6GLF, has just completed a 2,500-mile 2-meter QSO with Paul, KH6HME, in Hawaii while running 3 watts into a KB6KQ loop at a Southern California beach. (Photo by Gordon West, WB6NOA)

enhanced the most. For example, a signal path that's enhanced on 2 meters by tropo may not be quite as loud on 6 meters, or 432 MHz either. In fact, when going up in frequency, there may be more attenuation on the signals. There is no magic formula to determine which is the best frequency that can be supported by the duct, rather it is a trial and error process.

Indeed, because of the nuances and changing nature of tropo paths, the best time to work long distances is during one of the VHF contests that take place during the year. Tropo paths can spring up almost any time during the January, June, and September ARRL VHF contests. For example, Ken, WB2AMU, in grid FN30, has run 10 watts on 2 meters to work W4FSO in grid FM14 over 400 miles away during the June 2002 contest. It helps when both antennas are highly directional and pointing at each other. In Ken's case, he used an MFJ three-element Yagi and the CW mode (see **Photo 2-I**).

The time of day for best tropo ducting is not nearly as important as the building high-pressure system. In southern California, Gordo, WB6NOA, has worked tropo paths at high noon and well after midnight, and then watched the path drop out at sunrise and sundown. The next day, the absolute reverse might happen.

Paul, KH6HME, has discovered that tropo signals seem to remain strong to the West Coast as long as the blanket of clouds remains *below* his 8,000-foot elevation on the big island of Hawaii. Yet a year earlier, another operator on the same island, Al, KH6IAA, was heard loud and clear from his 40-watt mobile halfway up the side of the volcano in a steady drizzle. But he *did* indicate that his outside temperature was indeed warmer than when he left Hilo down below a few hours earlier.

Inversion tropospheric ducting exhibits such strong signal strengths that CW and single sideband may not necessarily be the *only* modes to work the path. Regular FM and even 6-MHz-wide, fast-scan amateur television contacts are possible over some extremely long paths. Five years ago, Gordo, WB6NOA, was able to receive a 6-MHz-wide, 100-watt amateur television signal from KH6HME over a 2,500 mile path between Hawaii and the West Coast. FM mobile stations regularly reach out to distant repeaters, well beyond line-of-sight range. In fact, VHF operator Pete Trotter, KB9SMG, confesses that he thought someone was pulling his leg when he worked into an ongoing repeater conversation while operating mobile on the expressway south of Chicago and was told the other stations were in Texas.

Beside looking up your local weather map and looking outside for an obvious inversion layer, *monitoring* the beacon subbands on VHF will tell you when the duct is beginning to form. For CW and single-sideband contacts, polarization is horizontal. This means a horizontal loop in the mobile, and a horizontal loop or horizontal beam at your house. The long boom beam is usually first to hear a distant CW beacon, but once that beacon builds in strong, operators with just a simple horizontal loop on their roof can join in on the reception. Once the beacon is coming in strong, operators begin to "make noise" on the national weak-signal calling frequencies using CW or single sideband. Now that many HF radios come with 6 meters, 2 meters, and 432-MHz SSB capabilities, weak-signal activity is at an all-time high.

Once you hook up with a distant station on CW or SSB, the practice is to move off the calling channel by at least 10 kHz. Move up or down, but find a spot for your QSO off the calling channels so other tropo stations can link up. A modest power

Photo 2-I. When tropo is working well it does not take much to work long distances! For example, Ken, WB2AMU, used this three-element Yagi, made by MFJ, at a height of 10 feet to work AA3ID in Cape Hatteras Island, North Carolina, with loud signals both ways. Again, the small wavelength of 2 meters makes a portable setup easy. AA3ID worked several stations in the New York area, including mobile and other stations with portable setups. (Photo by Ken Neubeck, WB2AMU)

level of 50 watts is plenty for SSB, and many sideband operators have worked tropo paths over 800 miles using as little as 2 or 3 watts into a beam fed with top-quality, low-loss coax.

On FM, tropo contacts are usually vertically polarized. FM repeaters are a great way to tell if a distant station is coming in via tropo. Tune to unused repeater channels in your area, and turn your VHF or UHF beam antenna to the expected direction of the incoming station. Aim it in the direction of the high-pressure system between you and the other operator. Tune around with the squelch off, and get set for some surprising reception.

Another trick is to tune into the 24-hour, 162.400 to 162.550 weather stations. Most weather stations regularly say the city of their local area broadcast, and when you hear a station indicating light winds along the Great Lakes—and you're in eastern Utah—you know that the band is preparing to open. Still have an outside television antenna? If so, look for distant TV stations by scrolling through unused channels and looking for some faint snowy pictures that ordinarily aren't there. TV stations pump out megawatts of effective radiated power (ERP), and many tropo experts may monitor specific TV channels for SSB activity to detect long-haul openings. Although cross-polarization may amount to as much as a 10- to 15-dB loss, a good powerful incoming tropo signal might still be detected.

Another important consideration when working tropo is to *always* say your city and state (or grid square). This is especially true if you have a number in your callsign that doesn't match up with your local location. Calling CQ with a 5 in your callsign from upstate New York may have Frank on Long Island mistakenly believing that the band has opened up to Texas again. He won't be happy to hear you're only 90 miles away and didn't give your approximate location.

Low-Pressure Tropo

Tropospheric ducting might also occur along incoming storm front walls of cold air colliding with warm air. This type of duct may sometimes extend well over 1,000 miles! It is not caused by subsidence in an inversion, but rather by a sharp change in the refractive index of air along the storm front wall. These low-pressure systems create a cold weather front, sometimes breaking apart with multiple, sharp-boundary cold/warm areas. They usually travel from northwest to southeast, typically curving north to give southern operators some exciting tropospheric ducting conditions to the northeast. A jet stream analysis may help you define an area that your VHF/UHF signal may skirt along for some exceptional, though short-lived, tropo distant contacts.

Unlike the subsidence inversion, the cold front tropo condition may only last for a few hours, and may require single sideband or CW to complete the contact. Your usually straight-line signal is bent along the face of the approaching storm front, sometimes giving you contacts of over 1,500 miles away.

Keep your eye on your barometer. Conditions will usually persist as the barometer stays steady. When it begins to drop as the cold front overtakes you, distant tropo signal strengths will also drop. According to Bruce, WA9NEW, off the East Coast, the physics of the fluid (of atmosphere) may produce the Bermuda High, and there is certainly good reason to believe that some of the infamous Bermuda Triangle aircraft/ship losses may well have resulted from the same kind of anomalous communications problems. He adds that upper altitude maps will show that this high or ridge extends to many thousands of feet through the troposphere and is a semi-permanent feature of the troposphere at a certain time of year. The basic laws of fluid dynamics dictate that air descends in such a region. Such descent to levels of increased pressure dictates warming. As air nears the surface, it must out flow, and the result is a boundary layer topped by a super-refractive layer.

Yet another interesting "trigger" of widespread tropospheric ducting is the formation of a hurricane to the south of the path that is beginning to form. Some tropo experts have written articles that claim that there is no proof that hurricanes to the south have any influence on more northerly paths. Yet ask any operator in Southern California, Hawaii, Texas, or Florida and they can just about predict when the band will open when a hurricane is spotted to the south.

Years of Pacific tropospheric ducting were documented by the late Robert Cooke, W6PJA, who regularly logged distant reception reports throughout the year, always growing in number with a hurricane located south of the path. Gordo, WB6NOA, also has an extensive collection of weather facsimile charts during strong band openings that always show the greatest signal strengths during the first two days of subtropical moist air overriding hot, smoggy air associated with subsidence inversion. Bingo, the band opens.

And, finally, there is almost always an open "path," via the troposphere, of invisible concentrations of compressed air masses, leading to tropospheric scatter. Two VHF stations separated by up to 500 miles may enjoy a QSO for hours on end via forward tropo scatter modes. Signals enter the upper atmosphere into a thin layer of super-refractivity, and travel well beyond line of sight in the same region that jet aircraft fly (30,000 to 40,000 feet up). It usually takes a minimum of 100 watts to sustain the path, and almost nightly, hams in Southern California talk on 2 meters and 70 centimeters to the San Francisco Bay area with S-4 signals that remain steady for months on end. Even during periods of rain and high winds down at ground level, the path remains open and requires only a modest beam and a minimum of 100 watts of SSB or CW power.

In Summary

In this book you will soon read about other exciting long-range propagation modes, but most of them require monster antennas and high-power amplifiers. For tropospheric ducting, however, modest Yagi antennas with no amplifiers may do the trick nicely. While we can't guarantee that you'll be able to tell exactly *when* a tropo path may open in your area, you now know that *everything* depends on your local weather and the "fluid of air" as it begins to stratify, like olive oil floating on water, when conditions become stable.

So when it gets too hot to do much of anything outdoors, drag out the morning paper, look for the stationary high-pressure system, and get set for some exciting tropospheric ducting conditions. And if there is a big storm front moving in tomorrow, check out that path that you might be able to work today!

References
1. Michael Mooney, "Sea Mirages Explained," *Ocean Navigator*, Landfall, September/October 1992.

CHAPTER 3

THE MAGIC OF SPORADIC-*E* PROPAGATION

One of the most significant modes of VHF propagation for most parts of the world is sporadic-*E* (abbreviated *E*s). This propagation mode appears initially at the higher HF frequencies, beginning around 28 MHz, and can affect the lower half of the VHF spectrum, reaching frequencies as high as 222 MHz on rare occasions. *E*s is a major propagation mode on the 6-meter band and is one of the key reasons we call it the "Magic Band." In this chapter, we'll learn how the mode was discovered, its importance to radio propagation, and the mechanisms behind its formation.

As the name suggests, *E*s propagation is not a regularly occurring propagation mode, and it is difficult to predict exactly when it may appear. At best, past observations can help us predict general occurrences, but only in terms of likely months (as opposed to days) that it may be encountered. It is truly a statistical-based phenomenon.

Of all of the VHF bands, *E*s appears most often on the 6-meter band, with an average number of occurrences in the United States and Canada between 70 and 100 days a year. This is because *E*s initially appears at lower frequencies and climbs up as the density of the formation becomes thicker. While it can show up on 2 meters, this is relatively rare in the United States and Canada (typically only once or twice a year). In other areas of the world, around Japan for instance, *E*s is not only a regular occurrence on 6 meters during the summer months, but can often reach the FM broadcast band and the 2-meter band as well.

When *E*s appears on a quiet 6-meter band, the effects are quite pronounced. Signals seemingly appear out of nowhere and will be quite strong. There may be rapid fading in and out of signals at times, going from weak to strong and back to weak again. The duration of an *E*s opening can be as short as a minute or as long as several hours. There is no way to predict how long an opening will last, so there's usually a sense of urgency in exchanging information and working as many stations as possible during an opening. *E*s on 6 meters is a fun activity for hams, particularly when using mobile and portable setups.

VHF Propagation Audio Course available. See Appendix II for details.

A typical *E*s opening on 6 meters will cover as far as 2,200 kilometers (1,350 miles). For example, Ken, WB2AMU, at his location on Long Island can work into the Ohio and Indiana areas on an opening that occurs to the west of him, and into Georgia and Florida when an opening occurs to the south of him. Gordon, WB6NOA, at his Southern California QTH can experience openings to the north enabling him to work the state of Washington, and to the east which opens states as far as Colorado and Wyoming. As a rule, there is generally no way to predict exactly what direction an opening will take or exactly what time of day it will appear.

The *E*s mode has been of intense interest for many decades, particularly as it occurs in the temperate zone. Temperate-zone *E*s appears consistently every year in both hemispheres, with a major season during the summer months and a lesser one during the winter. The peak is near the summer solstice for each hemisphere, with events appearing in over 60 percent of summer days.

The protocol for a 6-meter contact made through *E*s propagation is to exchange the important information (signal report, grid square, and name) up front because of the uncertainty of the opening's duration. However, on 2 meters *E*s openings tend to be very short, so a quick of exchange of grid square information is typically the rule of thumb when this happens.

What We Know

An *E*s formation is typically a thin layer (less than 2 kilometers thick), which is a subset of the overall *E* region. It can occur anywhere between 90 and 120 kilometers. It appears to have little correlation with the solar sunspot cycle, and while some variation between consecutive years may be observed, little of it can be directly linked to changes in the sunspot cycle.

Radio amateurs in the United States first discovered the phenomenon in 1935 on the 5-meter band (56 MHz), and it was through initial radio observations that a general idea was obtained concerning the behavior of *E*s propagation. **Figure 3-1** is a pictorial description of how radio waves are reflected off *E*s

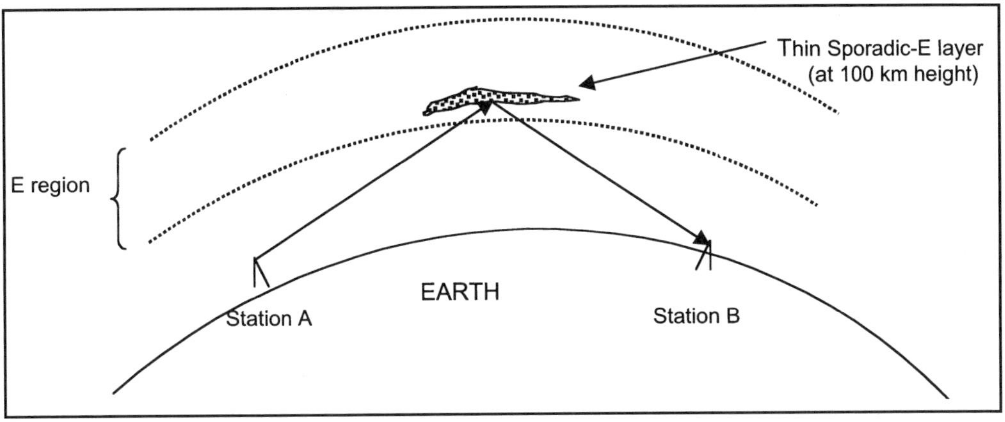

*Figure 3-1. The figure above shows how radio signals are reflected off a sporadic-*E *formation between two stations.*

clouds. The mode allows for long-distance communications to occur on frequencies beginning around 28 MHz up to 150 MHz. It can provide very efficient propagation during the summer months for the frequency range of 28 through 50 MHz for much of the world, and even higher in some areas.

Much of how and why the mode occurs remains a mystery, even half a century after its discovery. Since then, scientists and hams alike have studied *E*s through various means, with more knowledge being gained about this mode, but only at a gradual pace. We will discuss some of the scientific findings here as they relate to what radio amateurs observe on the VHF bands.

In 1940, scientists Appleton and Beynon[2] determined the relationship between the Maximum Usable Frequency (MUF) and the critical frequency (f_0) or the vertical frequency that is associated with *E*s propagation. They developed a formula for determining the critical frequency:

$$f_0 = \frac{f_{max}}{\sqrt{\dfrac{4\,(H + D^2/8R)^2 + D^2}{4\,(H + D^2\,8R)^2}}}$$

where:

f_{max} is the MUF in MHz (for which the transmission path is the skip distance)

H is the vertical height from earth to the cloud

D is the transmission path distance

R is the radius of the earth

(All distances are expressed in kilometers)

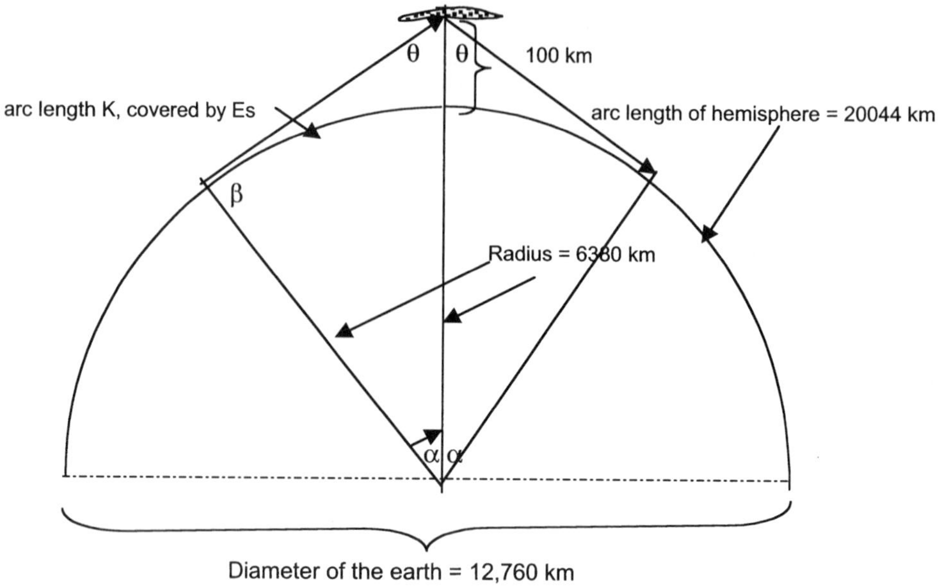

This example shows the general maximum distance that can be covered by a single-hop sporadic-*E* path. For this exercise, we assume that the entrance angle of the signal to the cloud (θ) is the same angle that is leaving the cloud. Using geometry, we can determine the angle and associated arc length by the two legs of the triangle (the radius of the earth to where the antenna is located and the radius plus the height of the sporadic-*E* formation), as follows,

$\cos \alpha = 6380 \text{ km} / (6380 \text{ km} + 100 \text{ km}) = 6380 \text{ km}/6480 \text{ km}$

$\alpha = 10.1$, $2\alpha = 20.2$ (this is also the arc length in degrees)

We can solve for K, the arc length covered by the sporadic-*E* skip, by setting up a ratio of the arc degrees with respect to arc length and using the arc length of the hemisphere (which is one half the circumference of the earth, or 20044 km),

$20.2/180 = K/20044 = 2250$ km (length of arc K covered by sporadic-*E* path)

Using geometry, we see that the angle of the signal leaving the antenna (β) is approximately 90 degrees.

Figure 3-2. Method for calculation of maximum distance covered by single-hop Es.

A rough approximation that is often used is:

$$MUF = f_{max} = 5.33 \times f_0$$

Figure 3-2 illustrates the method for calculating the maximum distance covered in one skip of *E*s in terms of the radius of the earth.

It is through these mathematical calculations developed by Appleton and Beynon that the height of *E*s formations was determined to be in the area of 100 kilometers ± 10 kilometers. However, scientific instruments were later developed that detected and measured formations in the ionosphere specifically at this height.

One way to measure the critical frequency and the height of the *E*s formations is through the use an ionosonde. Ionosonde stations send a range of radio waves up to 10 MHz directly vertical from the station and record the frequency of any radio waves reflected back from the different layers of the ionosphere. These stations were built and used in force from the 1950s up through the present and are located throughout the world. In the United States, a major ionosonde that's been consistently running is the one in Boulder, Colorado, at the NOAA facilities. **Figure 3-3** provides a pictorial description of an ionosonde station.

It is through the use of the ionosonde and the resulting ionograms (films of hourly reflections that show the shape of the formation) that different types of *E*s can be identified. These include the common flat type, cusp shape, and slant type. Sometimes knowing the type of *E*s formation is useful in deter-

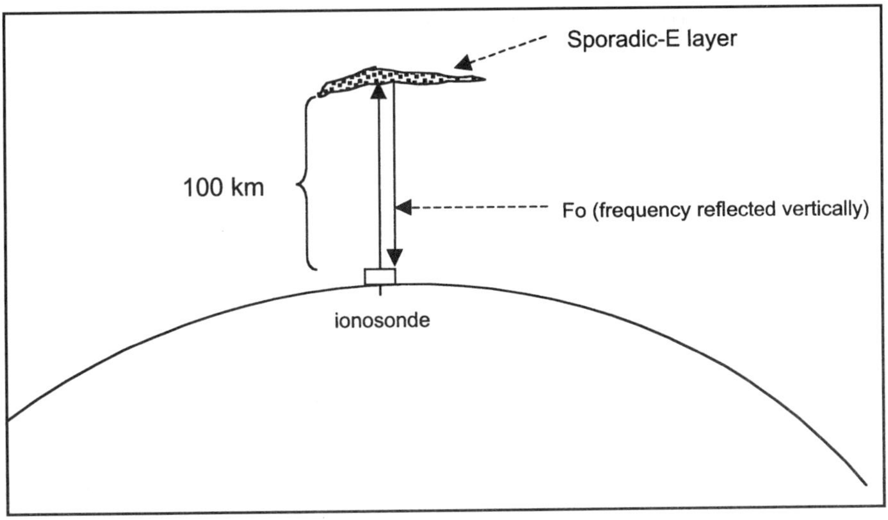

*Figure 3-3. A simple diagram showing the signal that is reflected off a sporadic-*E *layer back to an ionosonde.*

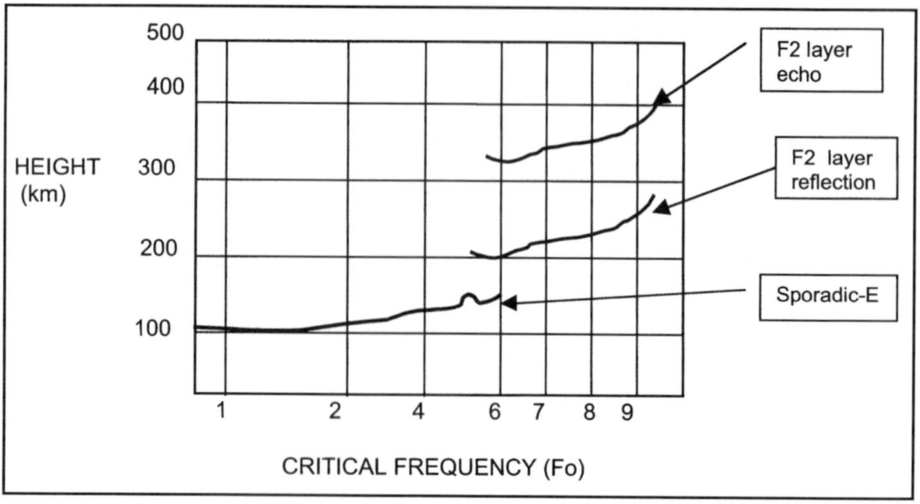

Figure 3-4. An ionogram shows the graphical results of reflections that come back from ionosonde frequency sweep soundings. From the figure above, we can see that the sporadic-E reflection reached a critical frequency (f_0) of 6 MHz, which translates to a formation that can reflect 10-meter signals.

mining how far signals may be transmitted. Analysis of the ionograms has also added to our knowledge of the phenomenon, placing its occurrence typically in the 85- to 115-kilometer height range. An example of an ionogram is shown in **Figure 3-4**. Note that the ionogram uses logarithmic scale in the X axis (frequency).

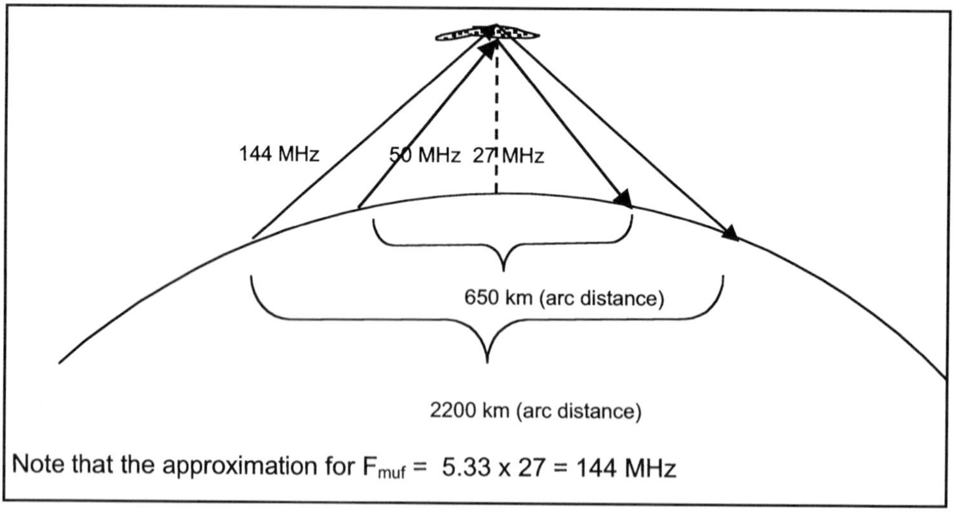

Figure 3-5. The figure above shows the relationship of 144-MHz and 50-MHz skip distances when there is a critical frequency (f_0) of 27 MHz.

In addition to the relationship between critical frequency and MUF, there is also a relationship between the MUF of different radio frequencies. This relationship not only applies to *E*s propagation but also to the F_2 propagation mode (detailed in **Chapter 6**). For example, if the *E*s cloud is reflecting very short distances with strong signals on the 50-MHz amateur radio band, on the order of less than 500 miles, there is an excellent chance that *E*s propagation may be present on the 144-MHz amateur radio band. This is because the density of the cloud has become thicker and reached a point where it can reflect higher-frequency radio waves.

In order for *E*s propagation to occur on 144 MHz, the critical frequency (f_0) has to be 27 MHz, with the distances between stations noticeably shortened on the 50-MHz band (see **Figure 3-5**). Thus an interesting mathematical aspect of this natural phenomenon is that higher frequencies support greater distances, while lower frequencies support shorter distances when the density of the cloud formation reaches sufficient levels.

The formula that shows the relationship between the MUF and the f_0 can be derived from the Appleton-Beynon equation as developed by Davies[3] as:

$$f_{0b} = fv\ 1\ \sqrt{(D/2h)^2}$$

where:

f_{ob} is the oblique frequency
fv is the vertical frequency

A rough approximation of this formula is derived from Emil Pocock, W3EP,[4] for *E*s formations at 105 kilometers. It is useful for most cases and is presented as:

$$d = 420\ \sqrt{(5.33\ f/f_{muf})^2 - 1}$$

For example, one can determine the distance between stations making contact on 50 MHz that would also indicate the possibility of an *E*s path being present on 144 MHz. Using the approximation formula, we get:

$$d = 420\ \sqrt{(5.33 \times 50/144)^2 - 1}$$

$$d = 420\ \sqrt{3.425 - 1} = 420 \times 1.56 = 655\ \text{km}$$

However, the distance calculated by these formulas does not mean that one cannot contact stations much farther away, particularly during the active summer period. Greater distances can be achieved when several *E*s clouds appear at the same time in a

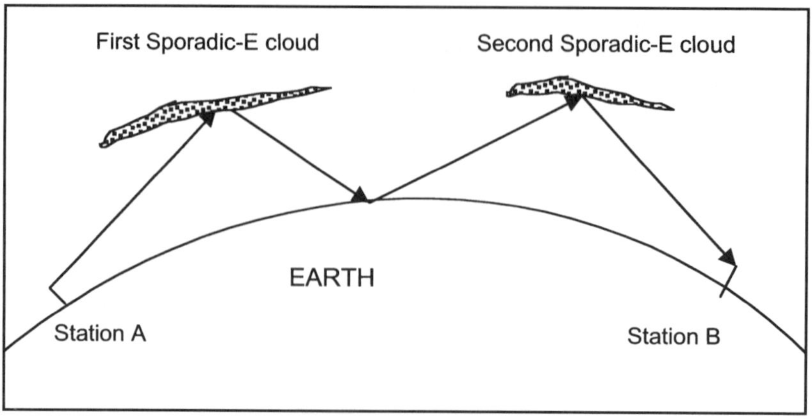

*Figure 3-6. The diagram above shows how double-hop sporadic-*E *signals are reflected off two sporadic-*E *formations positioned at the right locations. The overall distance covered by two sporadic-*E *clouds is on the order of 4,000 to 5,500 kilometers.*

hemisphere and enable "double-hop" and sometimes even "triple-hop" *E*s. This is how many amateur radio stations in the Northeast United States are able to contact stations on the West Coast or in Europe during the summer months on 50 MHz (see **Figure 3-6**).

Though relatively rare, triple-hop *E*s is possible during very active days in the summer. For instance, during a double-hop opening late in the afternoon into the United Kingdom on July 10, 2000, Ken, WB2AMU, was able to contact OK1DDO in the Czech republic on CW from Long Island. This was most likely a triple-hop *E*s opening (possibly even four hops!), and the contact was only possible through the CW mode because of the weak signal quality. Multiple hop conditions are not easily predictable and these events can happen any time, but peak periods for such events appear to be mid-morning and during the early evening hours of the summer.

An interesting exercise is to plot the general shape of the *E*s formation that occurs during a band opening on a geographic area map. For the most part, when a contact can be made via *E*s, the cloud formation is at the midpoint between the two stations engaged in the QSO. If several stations are heard on the 6-meter band during a short time period (say, 15 minutes to 30 minutes) from your location, you can plot a line from your location to each other location you can hear or work. By taking the midpoints of each of these lines, you can then determine the general shape of the formation.

Figure 3-7 shows the plots made using this midpoint method of some intense *E*s openings that Ken, WB2AMU, observed on 6 meters. As you can see, the formations can be quite large.

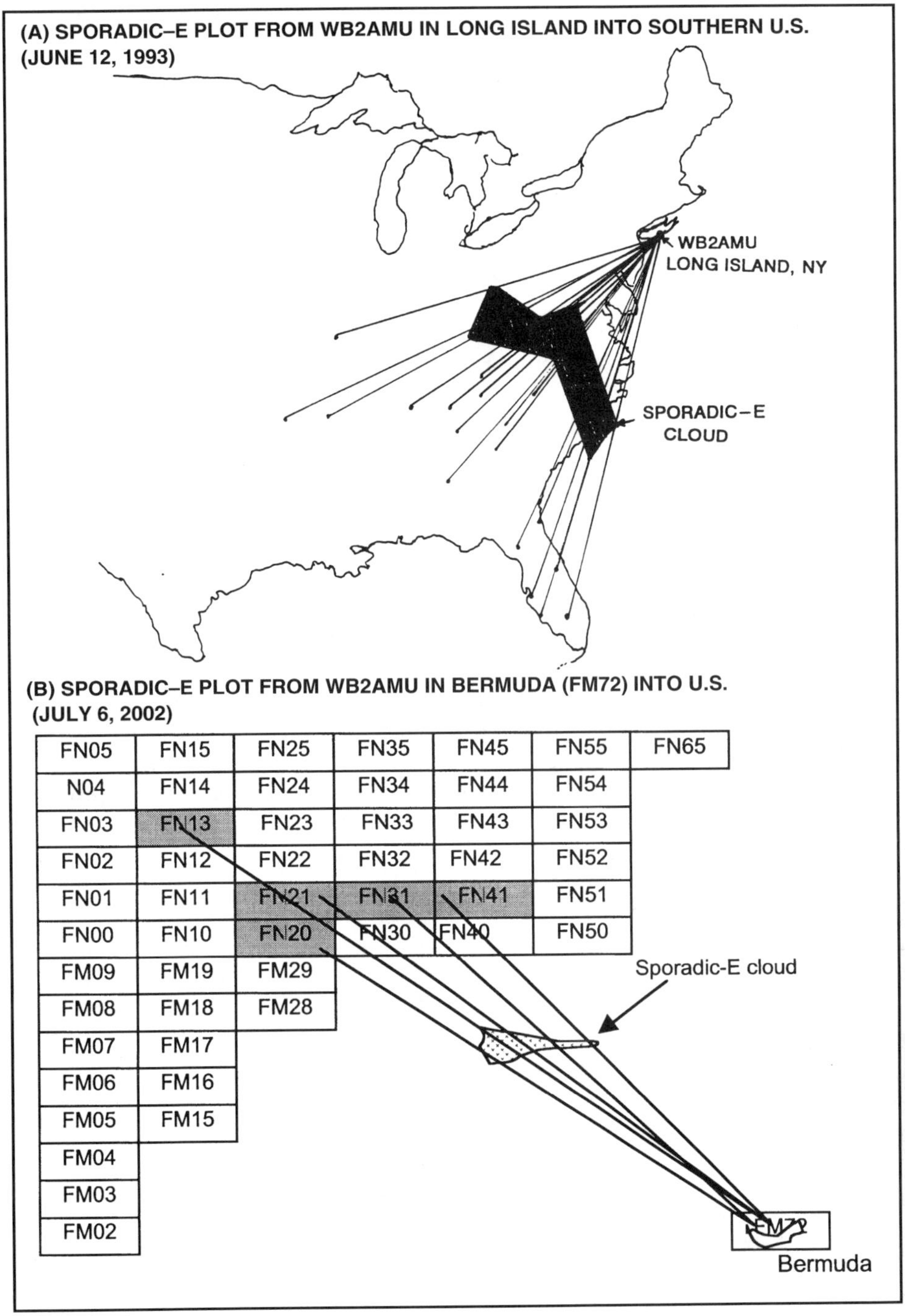

*Figure 3-7. In the figures above, we see that the rough outline of a sporadic-*E *formation can be determined by taking the midpoints of the paths of each contact made within a short time period. The plot in (A) represents a 30-minute opening, and the plot in (B) represents a 10-minute opening.*

However, other openings can be quite narrow and only one or two stations may be heard.

The *E*s clouds themselves will move in the *E* region. Experience suggests that the clouds tend to move in a westerly direction over time during the course of an opening, possibly as a result of the direction of the earth's rotation. The movements are typically gradual and slow.

What Sporadic-*E* Means for Hams

As *E*s is a very efficient form of radio propagation, signals can often be quite loud, regardless of the power level and antenna used. Indeed, on 50 MHz even a simple dipole antenna only 112 inches long or a quarterwave vertical 56 inches long with low power can be used to make contacts with other radio stations. Orientation of the antenna, with regard to horizontal or vertical polarization, is not critical as radio signals will change orientation when they travel through the ionosphere and are reflected off an *E*s formation. This effect of polarization change is known as Faraday rotation.

Of course, higher power levels and directional antennas, such as a two- or three-element Yagi beam, will help when conditions are marginal and provide a few decibels gain in signal strength. Sometimes the extra power comes in handy when the band is packed with signals, allowing you to break through the large number of stations. A higher gain antenna is especially helpful for QRP stations (10 watts or less on VHF).

When a strong, well-defined *E*s formation is present, there is also the opportunity to make radio contacts by means of the

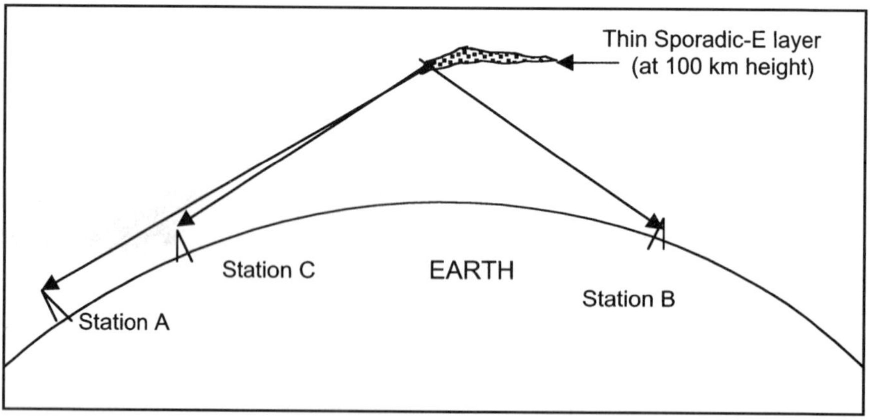

*Figure 3-8. This illustration shows how Station A is able to work station B through normal reflection off a sporadic-*E *cloud while at the same time is able to hear Station C, in the opposite direction, through the backscatter mode.*

backscatter method while facing the antenna towards the formation. Backscatter reflections typically come off the edge of a strong *E*s formation, thus stations located in the opposite direction of the *E*s cloud can be worked, sometimes as far as 200 to 300 miles away. An illustration of this situation occurred on July 10, 2000, when Ken, WB2AMU, was working many strong signals from eastern Canada while simultaneously hearing weaker signals from Maryland and Pennsylvania, all from his Long Island, New York, QTH (see **Figure 3-8**).

Why is *E*s propagation so important? The mode can actually be a double-edge sword, depending on the perspective of the user or service affected. For radio amateurs, it is the most important propagation mode on the 6-meter band. However, for commercial services (TV and FM broadcast radio) and public service radio users (police and firemen), *E*s skip conditions cause cross-channel interference from distant stations outside the preferred area of service.

This has resulted in great interest within the scientific community, in numerous studies being conducted, and also in frequency reallocations. In fact, in 1945 the FCC moved the original FM broadcast band in the United States from 42 to 50 MHz to 88 to 106 MHz. This move was made despite the objections of radio manufacturers and broadcast stations because of the expenses involved in changing equipment, but was necessitated primarily by *E*s causing cross-channel interference.

In hearings conducted by the FCC in 1945, it was noted during the sunspot minimum that there was much natural interference, primarily due to *E*s propagation. However, the 42- to 106-MHz range was set aside for the development of broadcast TV, so interference caused by *E*s remains an issue to this day.

Over 40 years' worth of ionosonde data has been collected, particularly at some of the longer running stations, such as the Boulder, Colorado, facility. Supplementing this data are the many thousands of radio amateur observations of this mode on the VHF band, particularly 50 MHz. Many veteran 6-meter operators (including the authors) have kept logs of all the stations they've heard and those they've contacted. This combination of data is extremely helpful in our understanding of the phenomenon, as documented in a paper by Neubeck, WB2AMU.[5]

Some of the more intense openings that have been observed have occurred during radio contests, such as the ARRL Field Day or the ARRL June VHF contest. An example of one of the most intense band openings ever observed by stations in the United States was during the ARRL Field Day exercises on June 25, 1998. Ken, WB2AMU, was operating at the Peconic Amateur Radio Club Field Day site in eastern Long Island using a modest setup of 150 watts and a three-element Yagi. There

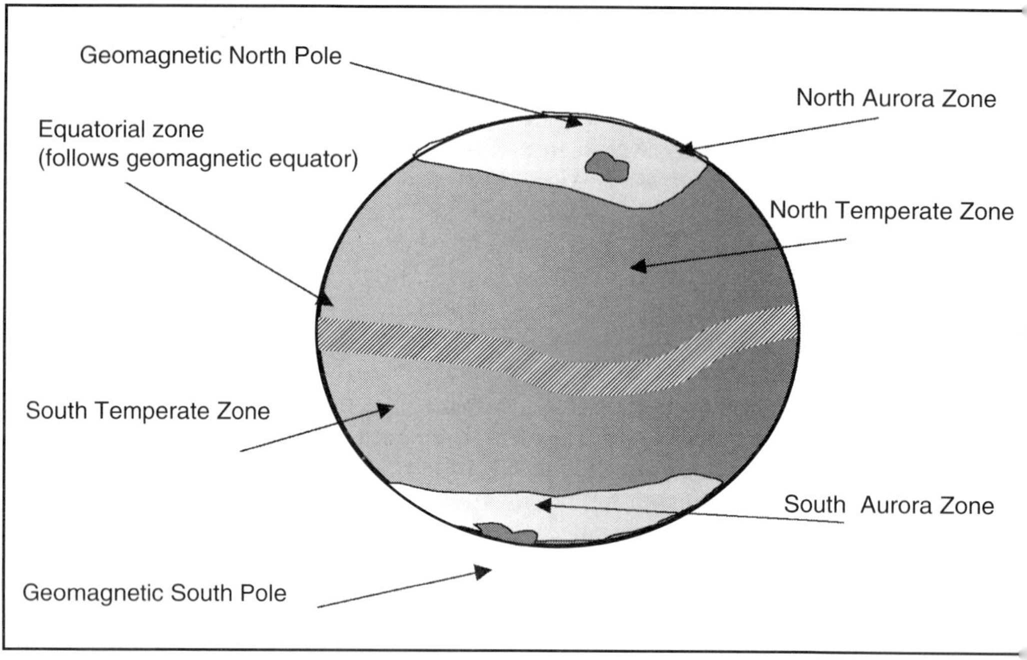

Figure 3-9. The figure above shows the various geomagnetic zones of the earth. Each of these unique zones has a bearing on different types of VHF radio propagation.

were *E*s openings throughout the entire 24-hour exercise, but the major event took place beginning at 8:00 a.m. local time on Sunday when a tremendous *E*s opening affected most of the United States. In addition to the single-hop openings, double-hop *E*s contacts were also made from this location into Colorado, Idaho, and Utah because of the presence of additional *E*s formations. Several radio stations located in the Midwest United States were able to contact stations in *all 48 contiguous U.S. states* during the 24-hour period because of the tremendous *E*s activity.

What Causes Sporadic-*E*?

So what is this mysterious phenomenon in physical terms and what causes it? Perhaps the easiest way to describe *E*s is as a formation of metal ions that are compressed into a thin stratified layer in the *E* region, located around 90 to 100 kilometers above the earth. The formation is capable of reflecting radio waves up to a certain frequency depending on its overall ion density.

The actual causes of *E*s have been the subject of much discussion and much research over the years. It's important to recognize that there have been many myths and unfounded speculation concerning the phenomenon, principally coming from radio amateurs. It's critical, therefore, to rely only on facts based on actual observations by radio and other scientific methods.

Radio amateurs have participated in a number of studies on the phenomenon, including a major 6-meter study in 1949 that was conducted by Oliver (Perry) Ferrell of *CQ* magazine[6]. Also during the 1940s, scientists, including Beynon and Appleton,[2] worked with hams and conducted analyses that related the critical frequency (or vertical frequency) to the MUF.

During the 1950s, *E*s was the focus of a major study conducted by Dr. Ernest K. Smith.[7] Besides collecting observational data based on radio and ionosonde findings, he described the three basic zones of the earth that shape the phenomenon of *E*s: the equatorial zone, the north and south auroral zones, and the north and south temperate zones (see **Figure 3-9**).

Within the equatorial zone, *E*s appears on a daily basis in the geomagnetic equatorial belt around the earth. The auroral zone sees aurora-type formations that scientists in the past may have mistakenly referred to as *E*s, instead of aurora propagation or auroral-*E* propagation. The phenomenon that's been of most interest, as well as the most challenging to define, is that which occurs in the temperate zone and has a unique annual pattern.

The basic morphology of temperate-zone *E*s is well known after years of hourly observations via ionosonde stations throughout the world, coupled with amateur radio frequency observations. The phenomenon has a unique annual pattern of a major peak during the summer solstice, a minor peak during the winter solstice, and two valleys during the equinoxes, with the spring equinox being the lower of the two. **Figure 3-10** shows the basic curve of this annual pattern and is based on several years of 6-meter data collected by Ken, WB2AMU, from his location in New York. It is similar to the patterns of ionosonde data collected at Boulder, Colorado.

This pattern is shifted by six months for stations in the Southern Hemisphere, with the maximum in the summer or December. The basic annual pattern appears relatively consistent throughout each year of a sunspot cycle, with perhaps a few more events observed during a cycle's quieter years. Data has also shown that *E*s experiences variations at points around the world. For example, the basic curve for 50 MHz in Japan would be even higher than that observed in the United States and Canada. Because of the unpredictability of *E*s, it's only possible to predict the number of occurrences on a monthly basis, not for specific dates or times.

A great deal of information was also gathered on the *E*s phenomenon through numerous rocket launches during the 1960s and 1970s. These were conducted from several locations over

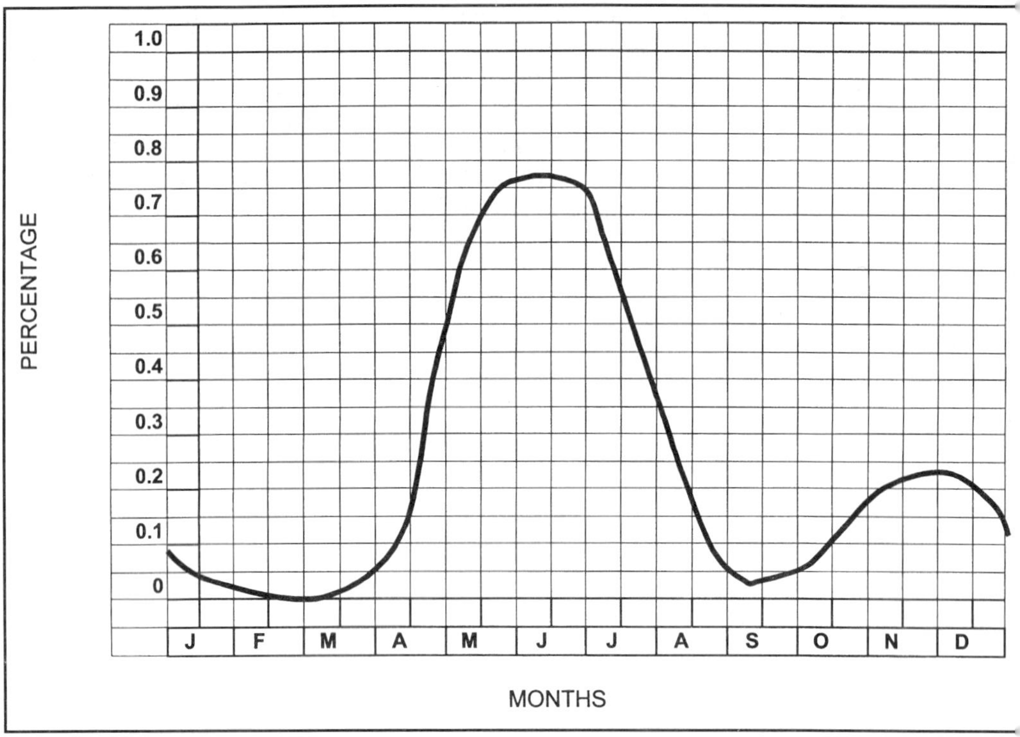

Figure 3-10. The figure above shows a very general curve of sporadic-E *occurrences based on observations made by Ken Neubeck, WB2AMU, and others in the northeast United States.*

earth and sent into the *E* region during an active *E*s opening. The on-board instruments were able to detect ionic current and electron density. **Figure 3-11** shows a typical example of the electron current data and the increased density at 100 kilometers. Another instrument, known as a mass spectrometer, was able to determine the atomic mass unit of the ions in the formation. Data collected from several launches consistently show that the ions that make up the *E*s layers are metallic, typically composed of iron, magnesium, sodium, or calcium. **Table 3-1** provides a summary of these launches into *E*s formations as well as other types, such as aurora (Au) and meteor scatter (Ms), and what ions were detected. Au and Ms data are also given in this chart to illustrate the similarities between these modes.

Table 3-1 also shows the consistent findings of ions of iron and magnesium at the same height, regardless of geographic location and time of year. The metallic particles found around 100 kilometers are presumed to be the remnants of meteor ablation in the *E* region. Recent research suggests that many of these particles stay in the *E* region for long periods, essentially creating a reservoir of metallic particles that can collect to form thin stratified layers when the conditions are right. Eventually many of these particles do fall from the *E* region through the process of oxidation at altitudes below 90 kilometers.

One of the principal mechanisms that help produce the thin *E*s layers at around 100 kilometers has been identified as wind shear (paper presented by Whitehead in 1988[8]). This is where winds from opposite directions compress the particles into thin layers. Gravitational forces like tidal waves are another important factor in *E*s and help to move the formations downward towards the 100-kilometer level (papers by Matsushita[9] in 1962 and Kirkwood[10] in 1997). **Figure 3-12** shows an actual *E*s formation moving along with a tidal wave to eventually dissipate below 100 kilometers. However, wind shear and tidal wave factors alone cannot fully explain the *E*s phenomenon, particularly its unique seasonal pattern.

Table 3-1. Summary of Rocket Launch Findings into *E* Region

Date	Location	Ions Detected	Height (km)	Formation
6/15/60	USSR	Magnesium, Iron, Silicon	103.5, 105	Es
10/31/62	Eglin AFB, Florida	Magnesium, Sodium, Calcium	85	Es
3/6/62	Ft. Churchill, Manitoba	Magnesium	100	Au
11/16/65	Eglin AFB, Florida	Iron, Magnesium, Calcium, Aluminum, Sodium, Silicon	94, 111	Ms
11/17/65	Eglin AFB, Florida	Iron, Magnesium, Calcium, Nickel	89, 93	Es, Ms
1/26/66	White Sands, NM	Magnesium, Sodium, Silicon, Iron, Aluminum, Calcium	106	Es
4/12/67	Eglin AFB, Florida	Magnesium, Iron, Silicon	93, 105, 115	Es
3/18/69	Heiss Island, Arctic	Magnesium	100	Au
4/10/69	Volgograd, USSR	Magnesium	105	Es
11/3/69	Ft. Churchill, Manitoba	Iron, Magnesium	107, 110	Au
11/20/70	Eglin AFB, Florida	Magnesium, Iron	95	Es
2//21/72	Heiss Island, Arctic	Magnesium	160	Au
7/3/72	El Arenosillo, Spain	Iron, Magnesium, Silicon, Nickel	113	Es
8/12/76	Wallops Island, Virginia	Sodium, Magnesium, Silicon, Aluminum, Calcium, Iron, Manganese, Titanium, Copper, Nickel, Potassium	101	Ms

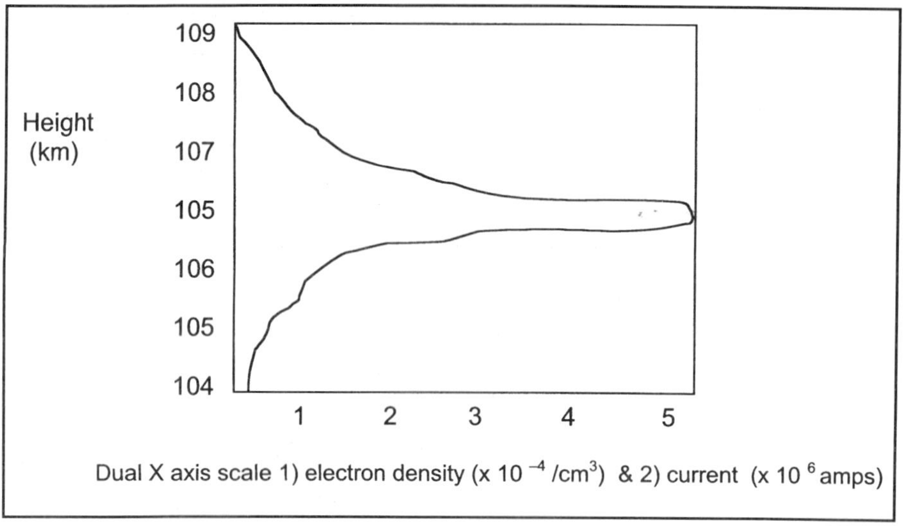

*Figure 3-11. This figure is typical of the scientific data that was collected during rocket launches, which were able to detect the formation of sporadic-*E *layers in the* E *region. In this example, a formation is detected at 105 kilometers.*

In recent years, observations using radar have been made which have provided additional clues to the movement of *E*s ions. In higher latitudes, upward-moving *E*s layers have been detected by the European Incoherent Scatter (EISCAT) radar setup at the Tromso and Sodrestrom facilities in northern Europe. Based on these observations, an empirical model describing the vertical transport of *E*s ions into the *F* layer and over the geomagnetic poles was developed by scientists Bedley and Watkins[11]. The model states that *E*s ions that are not neutralized at the 90-kilometer level (the top part of the *D* region) are transported upwards via prevailing ionospheric winds along with the geomagnetic field lines of the earth. Typically, these ions travel in a northeasterly direction towards the geomagnetic pole, reach the *F* layer over the pole, and then descend to the nighttime *E* region on the other side. The model assumes no impediment along the path of the ions as they travel towards the polar *F* layer.

Indeed this model for the higher latitudes could also explain a number of characteristics of temperate-zone *E*s occurrence. In the 80- to 100-kilometer region there are resident oxygen (O_2^+) and nitrogen oxide (NO^+) molecules. *E*s ions are "made" when the neutral (uncharged) metals from meteor ablation are ionized through a charge exchange or recombination process with these resident molecules. If these metal ions are in an atomic state, they can be re-ionized and lifted back into higher portions of the ionosphere to be part of the *E*-region metal ion reservoir. Those metal particles that remain neutral as well as those that become oxidized will eventually fall out of the ionosphere and towards earth.

The model shows why *Es* is more prevalent during the summer than it is in the winter months. At the 90-kilometer mark and below, the amount of atomic oxygen is lower in the winter due to lack of production by solar radiation. Thus during the

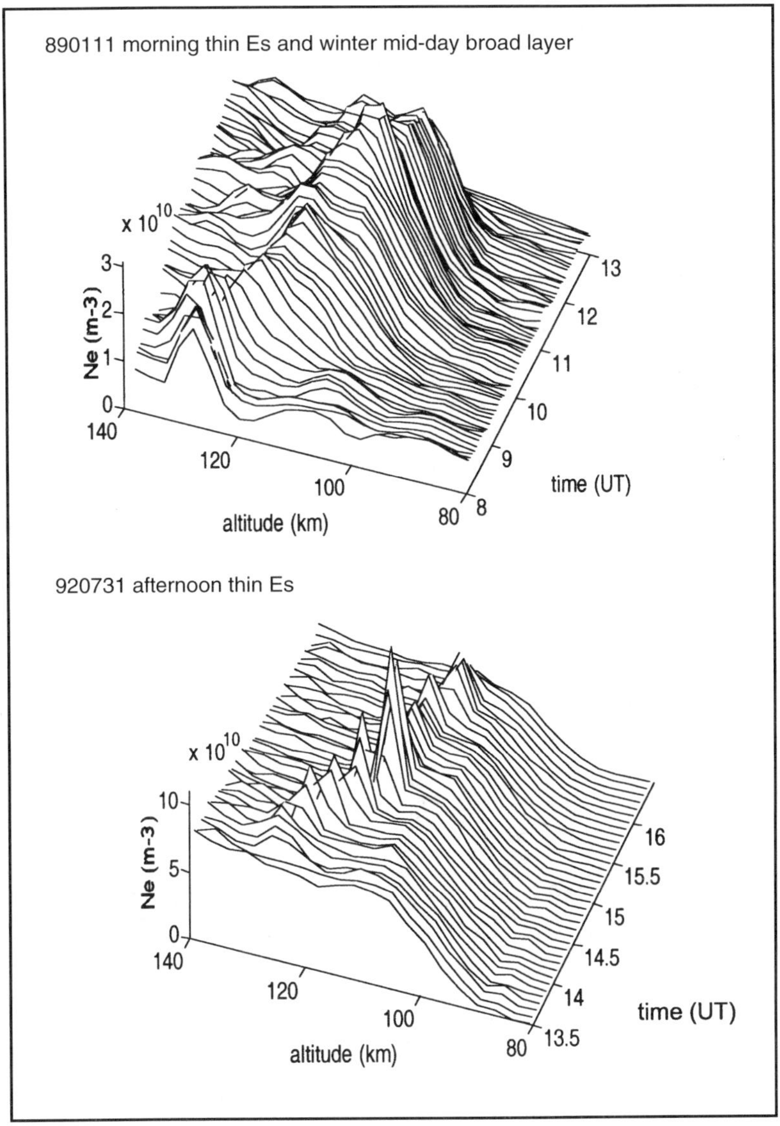

*Figure 3-12. The figures above were collected from EISCAT radar stations in northern Europe. They show both the thin layers of sporadic-*E *formation and its downward migration because of effects of tidal waves. As the formations reach below the 100-kilometer area, the formations start to break up. (Figures courtesy of Dr. S. Kirkwood)*

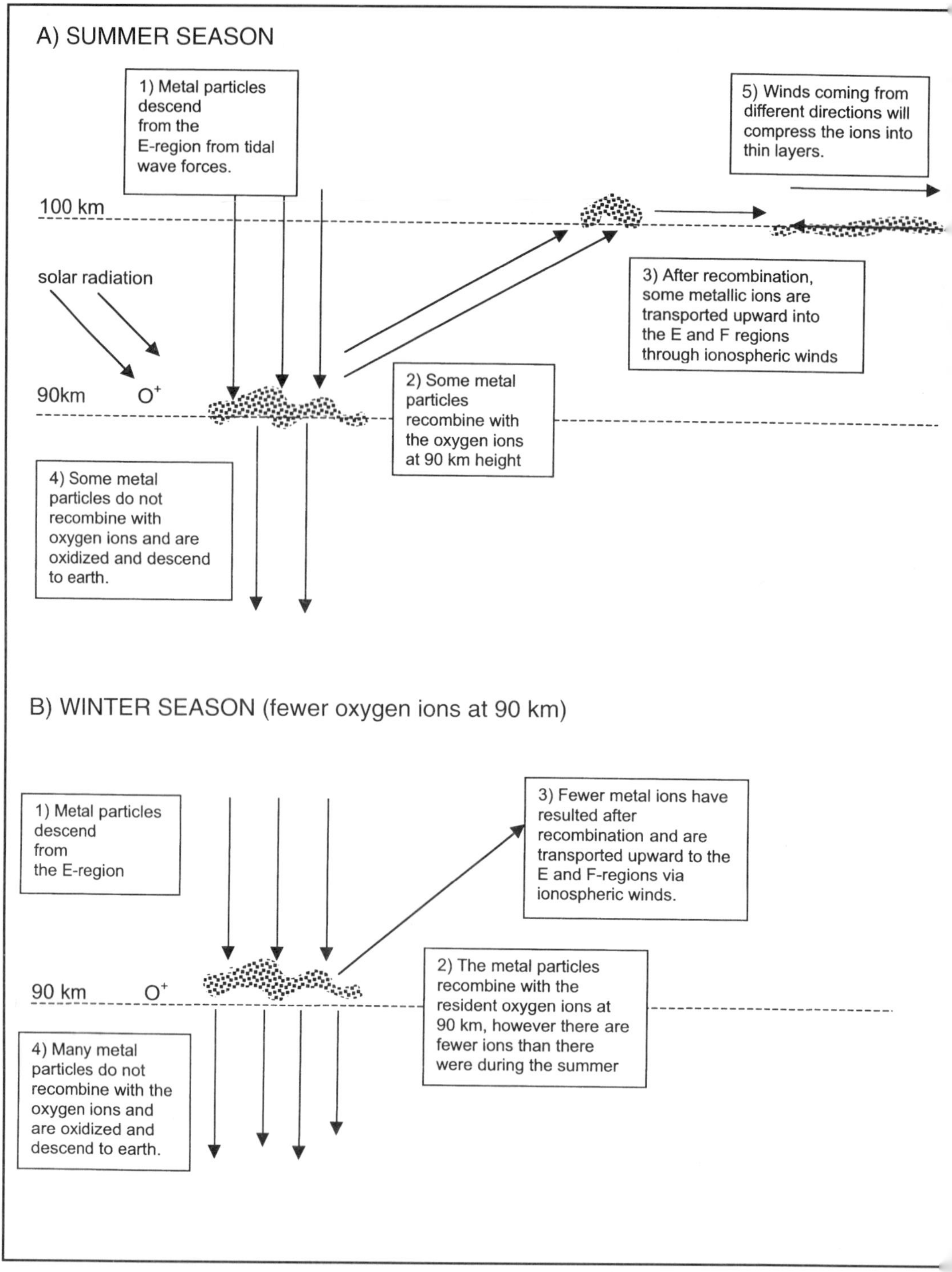

*Figure 3-13. The above figure represents a simplified explanation for the sporadic-*E *transport theory as explained in the text.*

winter, the neutralization or oxidation rate increases and fewer ions are formed through recombination which are available to be transported back into the *E*-region reservoir. Conversely, this reservoir of metal ions is predominant during the summer and is capable of providing ions for many *E*s formations (see **Figure 3-13**).

Along with the annual pattern, there is a diurnal, or daily, pattern to *E*s as well. It generally has two peaks during the day, one during mid-morning and the other during the early evening. It appears that there is a daily pattern of changes in the oxygen level at the 90-kilometer height that causes the two peaks of *E*s during summer days. These peaks appear to track with changes in the chemical processes of the ionosphere when the sun rises and sets, although there are also occasions when intense openings can last all day and after midnight. But, for the most part, *E*s is rare between 1 a.m. and 4 a.m. local time.

The *E*s phenomenon is certainly influenced by solar radiation, but the effect is an indirect one. It's based on the number of oxygen molecules present at 90 kilometers and changes during the course of the year. It also appears that, because of the unique physical makeup of the ionosphere at different heights, the creation of *E*s ions through recombination takes place more easily at 90 kilometers than at 100 kilometers. We can also see why *E*s is generally independent of the solar cycle, as the higher altitudes of the *F* layer comprise the area most affected by sunspot activity. This is in contrast to the annual solar radiation pattern effects on the oxygen level at 90 kilometers. The annual radiation pattern is consistent and shows no variance throughout the sunspot cycle.

The metal particles from meteor debris are not typically in an ionized state when first entering the ionosphere. Their ionization actually takes place at 90 kilometers and they travel upward to become part of the reservoir of metallic ions at 100 kilometers, in the *E* region. Scientists believe that this reservoir does not significantly change, even during intense meteor showers, as shown in a study by Baggedly and Steel.[12] This study showed that there were no trends of either overall decrease or increase of *E*s due to meteor shower activity.

It must be remembered that those meteor particles that are not burned and remain in the *E* region are in a neutral state. Thus a reservoir model for *E*s ions in the *E* region would explain the consistent nature of the phenomenon, where losses of metallic ions from the *E* region due to oxidation are compensated for by a daily influx of meteor particles entering it. This model fits the observations of hams and scientists regarding the symmetrical behavior between the Northern and Southern Hemispheres.

The empirical transport model described above appears to be supported by the radio amateurs' observations of late night *E*s openings on 6 meters at high latitudes during the summer months and occasionally during the winter months. This suggests that some *E*s formations are present in the aurora zone and may actually be at some stage of the transport process. These premises will ultimately be tested when oxygen levels are collected by future satellite launches.

Indeed, 6-meter observations have shown a sort of inertia effect that would support the transport theory. At the end of the summer season, there appear to be some "spillover" events of *E*s, indicating that some ions from the *E* region are still available for cloud formation. However, from the end of the winter up to about the beginning of the summer season (usually around May 1 in the Northern Hemisphere), there appear to be very few ions in the *E* region from which to draw. This may explain why the valley following the winter is more pronounced than that following the summer.

Stations located in the middle to high latitudes of North America should also remember that, prior to May, the *E* region is largely depleted of metallic ions because of the lower oxygen levels at 90 kilometers, and ions must build up again in the region for *E*s activity to begin. Thus a kind of "switch effect" is observed, where there is virtually no activity on 6 meters for many stations in higher latitudes during the month of April, followed by spotty activity during the first half of May, and then by regular activity during the second half of May and for the next few months. This effect has been observed for years by 6-meter operators in much of North America. Meanwhile, stations in the southern part of the United States start to see some *E*s during April as they enjoy the initial effects of increased amounts of sunlight (and solar radiation) before the higher latitudes do.

An interesting effect occurs at higher latitudes during the peak of the summer season, near the solstice. Stations at these latitudes see many hours of daylight and increased *E*s activity during the day. When Lefty, K1TOL, and Tim, NA1CW, went to Miquelon Island off the coast of Newfoundland at 46° North, near the summer solstice of 2002, they experienced over 18 hours of daylight and, consequently, would see *E*s activity soon after sunrise, as early as 7 a.m. local time. The higher latitudes certainly appear to be the place to be during the summer season for 6-meter *E*s!

While the model of the vertical transport of *E*s ions and the premise of seasonal change in oxygen levels explains a good deal of the physical mechanism of *E*s formation, further characterization is still required. Specifically, the key observation of a noticeable void of *E*s activity during the equinox months (March and September) in the temperate zones needs to be addressed. A reasonable explanation for the void can be hypothesized within the transport model through the combined resources of amateur radio observations on 50 MHz and detailed analysis of geomagnetic data.

*E*s occurrences in the temperate zones are very rare during the months of September, October, February, and March, regardless of zone. These are the months surrounding the fall and spring equinoxes. There is some thought that the higher geomagnetic activity that typically occurs during these months might be a contributing factor in inhibiting *E*s formation. It could be that Au formations that occur at the higher latitudes at this time may actually be blocking the transport of metallic ions that would otherwise develop into *E*s openings. (In the next chapter, there are examples of radar plots of Au formations that appear to be absorbing *E*s formations.) There is also some speculation that the alignment of the earth's magnetic field lines during the equinox period may also be an inhibiting factor for *E*s formation, perhaps impeding the process of the upward transport of ions.

Some major *E*s events on 6 meters have occurred on rare occasion during the equinox months and usually catch hams by surprise. For example, Ken, WB2AMU, has observed unusual *E*s openings on February 20 and March 19 in 1996, each event lasting more than two hours. The March opening was the first he'd ever seen during that month on 6 meters in several years of operating. It should be noted that 1996 was a quiet year in terms of geomagnetic activity, particularly during the first three months of that year, when the K index did not exceed the value 4.

An even more unusual event took place on February 20, 2001. On that day, openings between the northeast and southern United States took place beginning at 11 a.m. local time and lasted until 2 p.m. A second *E*s opening between the same areas took place later that day, beginning at 6:30 p.m. local time and lasting until 9:30 p.m. This particular opening resembled a summertime event (with a morning opening and an evening opening), yet obviously something very unusual took place in the *E* region, where there must have been a large influx of metallic ions causing the formation of strong *E*s clouds.

There are geographical variations within the general annual and diurnal patterns. For example, the highest incidence of *E*s occurs in the area of Japan, where the MUF typically exceeds 100 MHz during the summer months. Indeed, the phenomenon is a major problem because of the cross-channel interference encountered on the FM broadcast band and TV frequencies. On the opposite side of the world, South Africa experiences the fewest occurrences of *E*s, where even during the summer there may be less than a dozen openings. The variations between different areas appear to be related to geomagnetic factors of the earth, possibly concerning field lines and positioning to the geomagnetic poles. The United States has only moderate *E*s occurrences during the summer compared to some other areas of the world.

Sporadic-*E*: Friend or Foe?

Hams typically view *E*s propagation as a benevolent and highly desirable mode of propagation for certain radio services. As previously mentioned, however, many broadcast and public services view the phenomenon as the bane of their existence.

For example, a certain FM broadcast station in Japan operates on 90 MHz and, during the summer months, there is a high probability of *E*s propagation allowing other FM broadcast stations near the same frequency to cause interference, even from as far as 1,500 to 2,000 kilometers away. Complicating the issue is the fact that commercial stations use very high power levels. Thus the listening audience is getting interference from a distant station on the same frequency while listening for their usual local FM broadcast station.

What can be done in the case of *E*s interference? When this problem affected the commercial TV services in the United States, scientific groups conducted a number of studies on the phenomenon. However, there was little that could be done because of the allocation of many stations throughout the country on the same television frequency (usually Channels 2, 3, and 4 were most affected). The problem was significantly reduced only when cable TV became prevalent toward the end of the last century.

In the area of FM broadcast radio, however, there are more options for solving interference problems. FM signals are fairly wide, but they are still significantly smaller than the typical TV signal. In addition, a careful allocation could be made for each frequency assigned to a broadcast FM station, and a careful mapping of the locations of all stations in a major region would be a major step in correcting interference.

For hams, though, *E*s remains the key mode for making long-range contacts and is a major phenomenon in the VHF range, primarily on 6 meters and occasionally on 2 meters.

Summary

In this chapter, we discussed the known physical properties of *E*s propagation and its possible causes. We presented a working model presented by scientists studying the phenomenon, which also fits with many of the ham radio observations on the VHF bands.

Certainly, there are many things we still don't understand about *E*s, Some of the mysteries that remain for further research to explore are:

• Why does it behave as it does in the temperate zones?

• Why is there an absence of *E*s activity during the equinox months (March and September)?

• Why is there very little year-round *E*s in South Africa but tremendous activity in Japan?

• What series of events in the ionosphere causes singular openings during March and September?

The answers to these and other questions may be provided by the continued observations of radio amateurs on the VHF bands, in particular on 6 meters. Indeed there's still so much we don't understand about *E*s, and it seems that it may remain that way until more advanced capabilities are developed to gather data and additional observations are made that will capture all its nuances. But, until then, perhaps it's not such a bad thing to have a few mysteries remaining in our world to challenge us.

The sporadic-*E* phenomenon has been the subject of many myths over the years, but with the findings of scientists, coupled with amateur radio observations, the truth about its behavior is emerging.

References

2. Appleton and Beynon, "The Application of Ionospheric Data to Radio Communication Problems," *Proc. Phys. Soc.*, Vol 52, July 1940, Page 5183.
3. Kenneth Davies, *Ionospheric Radio Propagation*, Dover, 1964.
4. Emil Pocock, W3EP, *Beyond Line of Sight*, ARRL, 1992.
5. Ken Neubeck, WB2AMU, "Using the Combined Resources of Amateur Radio Observations and Ionosonde Data in the Study of Temperate Zone Sporadic-*E*," *Journal of Atmospheric and Terrestrial Physics*, Vol 58, No. 12, 1996.
6. Oliver P. Ferrell, "The Radio Amateur and Upper Atmosphere Research," *CQ* magazine, Cowan Publishing, February 1949.
7. Ernest K. Smith, "Worldwide Occurrence of Sporadic-E," *National Bureau of Standards Circular* 582, 1957.
8. Whitehead, "Recent Work on Mid-latitude and Equatcrial Sporadic-E," *Journal of Atmospheric and Terrestrial Physics*, Vol 51, 1988.
9. Matsushita, "Lunar Tide Variations of Sporadic-E," *Ionospheric Sporadic-E*, Pergamon, 1962.
10. S. Kirkwood, "Thin Ion Layers in the High-latitude Lower Ionosphere,", *Adv. Space Res.* Vol 19, No. 1, 1997.
11. D.R. Bedley and B.J. Watkins, "Large-scale Transport of Metallic Ions and the Occurrence of Thin Ion Layers in the Polar Ionosphere," *Journal of Geophysical Research* Vol 102, 1997.
12. W.J. Baggedly and D.I. Steel, "The Seasonal Structures of Ionosonde Es Parameters and Meteoroid Deposition Rates," *Planet Space Science* 32, 1984.

CHAPTER 4

UNDER THE SHADOW OF THE AURORA

Perhaps the strangest of all propagation modes that appear on 6 meters, 2 meters, and, on rare occasions, 432 MHz is aurora (abbreviated Au). The aurora borealis is one of our planet's most interesting natural phenomena. More commonly known as the northern lights, its displays of green and red formations often appear in the skies of the higher latitudes (see **Photo 4-A**).

As many VHF operators know, however, the aurora is more than just a visible phenomenon: it's also a significant form of

Photo 4-A. Aurora is another of the visual VHF propagation modes. Even in black and white, the intensity of the aurora formation can be seen in the skies. (Photo by Gordon West, WB6NOA)

radio propagation that takes place in the *E* region of the ionosphere. To the delight of many VHF operators, radio signals can be propagated in this mode by being reflected off the aurora in a backscatter manner.

As with sporadic-*E* (*E*s), radio amateurs were the first group of users to discover this mode of propagation, back in the 1930s on the 5-meter band. One of the earliest reports of an aurora opening was made by a ham in New York, W2AMJ, and printed in the May 1939 issue of *QST*. His report illustrates that the mode was then somewhat of a curiosity, as follows:

> The Northern Lights were going to town on the night of February 24th. I got on 56 Mc. At about 8:20 PM. I sent a CQ on i.c.w and was rewarded with a call from W8VO in Akron, Ohio, also on i.c.w. His carrier was S9 but had the most peculiar sound to it, with or without modulation. He told me that the band had been open out there for about an hour. I swung over to phone, and he advised me that phone was unintelligible. Then he shifted to phone, and boy, if you ever heard inverted speech you should have heard his. It sounded nothing like voice modulation. We continued the contact on i.c.w, S9 both ways. At no time was there any appreciable fading. The signal was just like it was around the corner, except for the peculiar combination of howl and roar that accompanied his carrier.

In 1946, 6 meters was allocated to hams (5 meters was discontinued at the beginning of World War II) and aurora conditions would be heard there also. Six-meter stations in certain locations would hear this weird-sounding distortion and know that an aurora opening was taking place. With their observations of radio aurora on 2 meters and 6 meters, ham radio operators would be very helpful to scientists studying the phenomenon. From 1957 through 1963, hams in Western Europe recorded their observations of aurora on 2 meters as part of a major study that was conducted and published as a paper by L. Hesse.[13]

Aurora and Amateur Radio

Aurora takes place in the *E* region, typically at 100 kilometers (about 70 miles) above earth, and it has the ability to reflect radio waves on different ham bands, depending on the density of the formation. However, in contrast to the forward reflections of *E*s, for many stations in the temperate zone (below the aurora zone), aurora is a backscatter or bistatic mode.

As aurora is centered around the geomagnetic poles (north and south), there are two auroral zones. During times of high geomagnetic activity, the aurora will extend into the lower latitudes, allowing stations just below this extended zone to work other stations via the backscatter mode. During auroral conditions, stations should point their antennas toward the nearest Pole to get

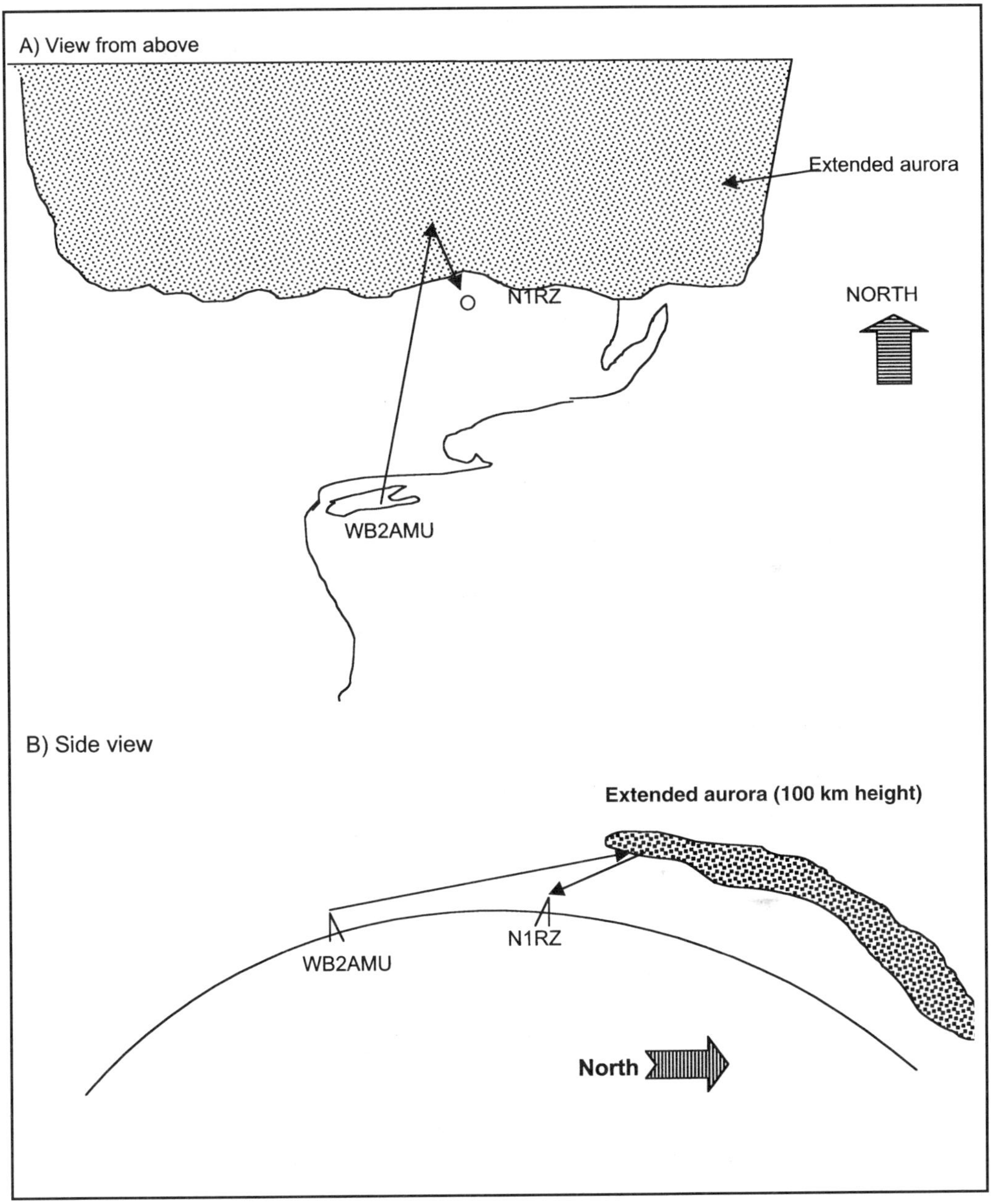

Figure 4-1. In this figure, WB2AMU from Long Island (FN30) is able to work N1RZ in New Hampshire (FN44) during an aurora opening on March 31, 2001, on 6 meters by means of backscatter. Both stations point their antennas north and are able to hear each other's CW signal. WB2AMU, using only 20 watts and a dipole, would not have been able to work N1RZ via direct path, even though the direct path (250 miles) was shorter in distance than the backscatter path (over 400 miles).

the optimal effects of backscatter. Specifically, stations located in the Northern Hemisphere should point their antennas north for their signals to reflect off the aurora and head back south.

Figure 4-1 shows a typical Au backscatter contact made by Ken, WB2AMU, from Long Island on 6 meters with another station in New Hampshire in March of 2001.

Like the *E*s phenomenon, Au propagation has also been of intense interest to scientists as well as hams. Au propagation in the upper portion of the temperate zones and inside the aurora zone appears more often around the equinoxes (September, October, March, and April). However, aurora events can occur at other times of the year, even the middle of the summer, if there is major geomagnetic activity present.

There are also differences between Au formations and *E*s formations. For instance, aurora is a broader layer, more than 2 kilometers thick, which can occur anywhere between 90 and 120 kilometers. The incidence of aurora openings also seems to track with the solar sunspot cycle, with more events occurring during the peak years of solar activity and for years following the peak.

As described in our quote from *QST*, auroral signals are severely distorted. SSB signals sound like voices gargling in water. One might even think that the receiver is breaking down! Most experienced 6-meter operators will agree that CW is generally the best mode to use during an aurora opening on six or the other VHF bands since voice modes tend to be very difficult to understand.

When going up in frequency during an aurora event, the frequency shift and Doppler effect is greater, and signals on 2 meters and above are greater than 1 kHz wide. Voice modes are, therefore, rendered almost unusable at this range. For those interested in working this mode, this is reason enough to know the code. If you've ever been in the position of trying to decipher the callsign of a horribly distorted SSB signal during an aurora opening, you know what we mean!

When giving an RST signal report to other stations involved in an aurora QSO, the convention is not to give a tone report because the tone quality of an aurora signal is severely distorted and not pure. We use the letter "A" in lieu of a tone report (typically number 9). A typical report would be 57A for an aurora-generated signal. The same report format is used for grading the signal strength of both CW and SSB signals, with a typical SSB report exchange being "57 aurora."

For those stations in the auroral zone and in the higher latitudes of the temperate zones, the best times for Au propagation

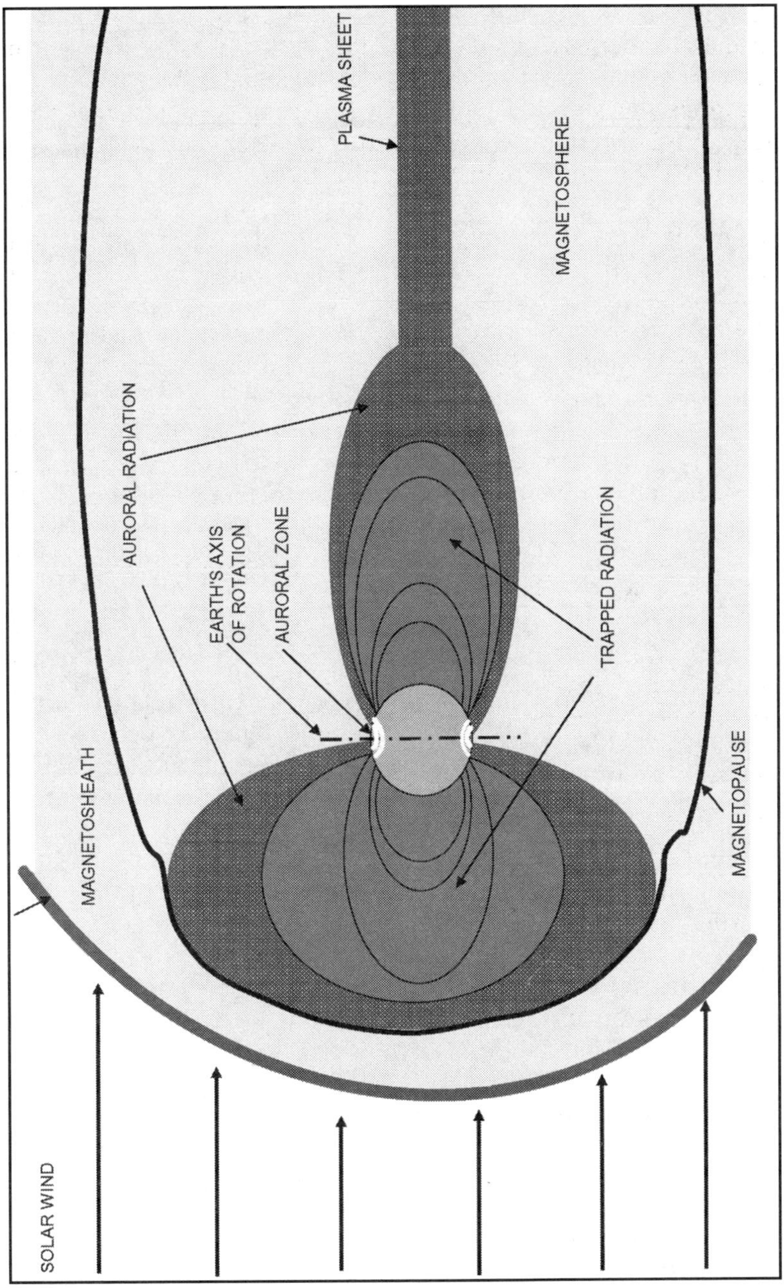

Figure 4-2. In this figure, the plane of the earth's axis is consistent with the position it is in during the fall and spring equinoxes. The earth is in a more favorable position, with regard to its axis, to receive solar particle precipitation that will affect the auroral zones surrounding the North and South geomagnetic poles. At other times of the year, the axis of the earth is not in the same plane (it has a 23.5-degree tilt toward or away from the sun) and is thus in a less favorable position for solar precipitation entering the auroral zones.

Table 4-1. K Index Values and Associated Effects on Radio Propagation

K index value	Geomagnetic Descriptor	Average occurrence per solar cycle	Typical effects on radio frequency communications
9	Extreme storm	4 days	HF propagation impossible, VHF aurora propagation
8	Severe	60 days	Sporadic HF propagation, VHF aurora propagation
7	Strong	130 days	Sporadic HF communications, VHF aurora propagation
6	Moderate	360 days	HF propagation fades at high frequency, some VHF aurora
5	Minor storm	900 days	short periods of VHF aurora propagation possible
4	Active	>1000 days	Generally no effect on HF propagation
3	Active	> 10,000 days	Generally no effect on HF propagation
2	Unsettled	> 10,000 days	Generally no effect on HF propagation
1	Quiet	> 10,000 days	Generally no effect on HF propagation
0	Quiet	> 10,000 days	Generally no effect on HF propagation

are generally during the fall and spring equinoxes. This is because during these periods the positioning of the earth relative to the active zones of the sun (solar eruptions generally occur in the areas just above or just below the sun's equator) is most favorable. These solar eruptions result in a stream of particles being ejected from the sun towards earth via solar wind at very high speeds.

Figure 4-2 is a graphical representation of this and shows the particle streams approaching the geomagnetic poles of the earth. During the equinox, the earth is positioned so that the plane for its axis of rotation is parallel to the plane of the axis of rotation of the sun. As you many recall from physics, the earth rotates in a fixed two-dimensional plane around sun.

It takes roughly two days for these particles to reach the earth's magnetosphere. There the electron precipitation interacts with the geomagnetic field lines and ambient particles in the *E* region at approximately 100 kilometers. This interaction is a complex process that can result in phenomena such as aurora formations and the flow of high amounts of electric current.

Geomagnetic activity can be measured at a specific location through the use of magnetometers and graded through the use of a K index, which ranges in value from 0 to 9, with the higher number representing the highest amount of geomagnetic activity. The K index is used worldwide. In the United States, K value measurements are taken by NOAA at the Boulder, Colorado, facility already mentioned. K index levels of 4 to 5 indicate a minor geomagnetic storm and could indicate that the aurora extends from Canada into the northern United States. Typically, a geomagnetic storm that wipes out most skywave propagation on HF can be on the order of K values exceeding 6.

While conventional *F* skip HF communications will be impaired (see **Chapter 6**), the potential of Au propagation on VHF bands increases during geomagnetic storms. Therefore, when major solar flares are observed on the sun, experienced hams start monitoring the VHF bands two days later. Refer to **Table 4-1** for a summary of K index values and their associated effects on radio propagation.

The color of the aurora indicates what elements comprise the formation. There are usually two basic colors for the aurora: green (Type C) and red (Type D). Other variations include green aurora with red outlining the top (Type A) or green aurora with read outlining the bottom (Type B). The color of the aurora is the result of its specific element composition as well as the excitation levels of its molecules. For example, the green aurora looks green because oxygen atoms are excited to a certain level; red aurora results from oxygen atoms being excited to a different level, where more red light at 6300 angstroms is visible to the naked eye.

In addition to oxygen and nitrogen gasses, there is evidence of active metallic ions residing inside the aurora, which may add to the radio wave reflection capability. Numerous scientific studies, particularly during the 1960s and 1970s, examined active aurora with instrumentation launched on rockets. (**Table 3-1** outlined results of some of these efforts.) Aurora studies were conducted in the high latitudes of the Northern Hemisphere inside the aurora zone at locations such as Poker Flats, Alaska, and Fort Churchill, Manitoba (see **Photos 4-B** and **4-C**).

In one such study, a rocket containing a mass spectrometer was launched in March of 1965 from Fort Churchill into an active aurora. The mass spectrometer recorded ion composition and the data was sent back to a ground station by telemetry. On this particular flight, a stratified layer of metallic Magnesium ions (Mg^+) were detected at 100 kilometers. This may suggest that these metal ions, resident in the *E* region, were excited by the electron precipitation caused by solar events, resulting in a radio aurora. It would appear that either the metal ions or the excited oxygen ions, or both, contribute to radio aurora.

Perhaps the most vivid view of the aurora was seen from space rather than from earth. Astronaut Owen Garriott, W5LFL, observed a major aurora formation over the South Pole while he was in the *Skylab* station above the earth in September, 1973. On September 7, *Skylab* astronauts witnessed a spectacular solar flare that also had a tremendous burst of x-ray flux that emanated from the violent solar eruption. About two and a half days later, particles ejected from the sun during the flare began arriv-

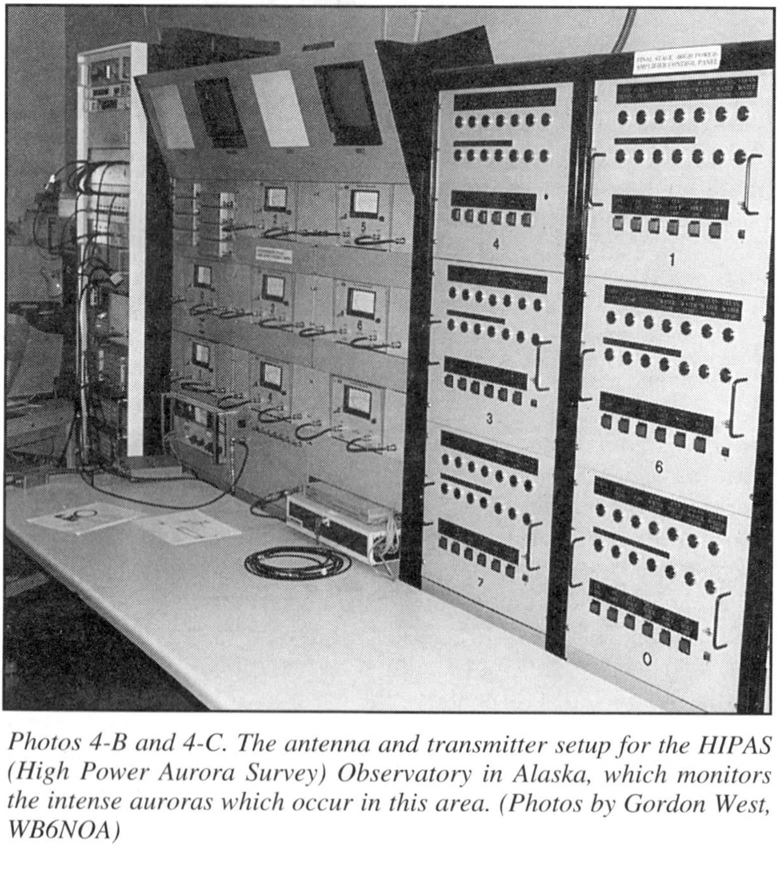

Photos 4-B and 4-C. The antenna and transmitter setup for the HIPAS (High Power Aurora Survey) Observatory in Alaska, which monitors the intense auroras which occur in this area. (Photos by Gordon West, WB6NOA)

ing at the earth's upper atmosphere. At the time, *Skylab* was in an orbit inclined only 51 degrees to the equator and the aurora was observed from the equatorial side, with Dr. Garriott taking several pictures with a handheld camera. On September 9, the Kp index reached a high value of 8—essentially storm level!

When there is intense aurora activity caused by major geomagnetic storms, the aurora zone can extend further down into the higher latitudes of the temperate zone, creating rare opportunities for hams in the southern states to get in on the fun. During some of the intense aurora openings in Cycle 23, hams as far south as South Carolina and Texas were able to complete aurora contacts via the backscatter mode!

The effects of geomagnetic storms can extend beyond radio propagation and visual displays to the point of causing major damage to power and telephone lines. This is because there are high levels of electrical currents flowing in the atmosphere during these storms, and they can reach as low as ground level, causing electrical shorts to power lines. One of the most intense geomagnetic storms in recent history happened during the first week of August, 1972, when three gigantic solar flares ejected matter that accelerated to speeds of almost 2.7 million miles per hour! The *Pioneer 9* spacecraft was the first to feel the shock. Beginning on the night of August 3, a series of brilliant aurora patterns radiated over much of North America and Europe. Power lines surged with voltage fluctuations, and a 230,000-volt transformer was blown in British Columbia. A similar magnetic storm took place some years later, causing electrical disturbances throughout October, 1989, and cutting power to millions of Canadians.

What causes the Doppler shift in the aurora signal? It's a complex process but, simply put, the mass movement of aurora *E*-layer electrons in the same direction relative to the earth creates powerful ionospheric currents. These currents propel the electrons at velocities of up to 3,000 meters per second, parallel to the earth's geomagnetic latitudes. These velocities are sufficient to cause major frequency shifts on VHF signals and, subsequently, a Doppler shift.

The unique sound of the aurora buzz is the result of a more complex process and has been the subject of many scientific studies. There are apparently many scattering regions inside an aurora, as the formation itself is not homogenous throughout. The random motion of electrons is believed to be responsible for this non-uniformity, applying a different sort of relative motion to the scattered radio signals that enter the formation. Thus the frequency shift for the Doppler effect and the auroral buzz may be caused by two separate sources of motion inside the aurora. Quite simply, there is a lot of stuff going on inside an aurora that affects the tonal quality of the radio signal.

Photo 4-D. It is not necessary to have a high-power station and super antenna setup to work VHF via the aurora mode. WB2AMU has made numerous portable excursions using moderate power and simple beams to work stations via aurora on 2 and 6 meters. In this photo, he is using a QRP portable setup on the north shore of Long Island, running 10 watts into a three-element Yagi mounted on mast sections up 15 feet high. Ken tapped into an aurora opening during the January 1998 VHF contest and worked K2AXX in upstate New York by simply aiming his beam north. (Photo by Ken Neubeck, WB2AMU)

One of the neat things about aurora backscatter on 6 meters is that it allows one to work many of the close-in grid squares that are beyond line-of-sight communications or tropospheric enhancement paths. For example, via aurora Ken, WB2AMU, can work into grid FN34 in upstate New York, which is too far from his grid of FN30 to work via line of sight and too short for typical distances covered by *E*s skip (see **Photo 4-D**). Distances of up to 1,500 miles (as measured by direct path) can be worked almost as successfully as with a single-hop *E*s opening. As in *E*s, the distances are limited by the actual height of the formation, which is typically in the 100-kilometer range.

An aurora opening becomes more intense where the density of the formation is thicker, and higher frequencies can then be reflected. Aurora can actually be heard on lower frequencies, such as 10 and 15 meters, but it's often noticed more at the beginning of the VHF range at 6 meters. After 6 meters, hams should start looking to 2 meters for activity. Aurora backscatter can allow for long-range 2-meter QSOs that would normally be impossible. However, at this high frequency, CW is definitely the mode to go with since signals are wider and more distorted.

One such intense opening occurred on March 31, 2001, when Ken, WB2AMU, was able to use moderate power levels of 50 watts on a portable setup in his driveway. He operated on 6 and

1) GRIDS HEARD OR WORKED ON SIX METERS BY WB2AMU (FN30)

EN65	EN75	EN85	EN95	FN05	FN15	FN25	FN35	FN45	FN55
EN64	EN74	EN84	EN94	FN04	FN14	FN24	FN34	FN44	FN54
EN63	EN73	EN83	EN93	FN03	FN13	FN23	FN33	FN43	FN53
EN62	EN72	EN82	EN92	FN02	FN12	FN22	FN32	FN42	FN52
EN61	EN71	EN81	EN91	FN01	FN11	FN21	FN31	FN41	FN51
EN60	EN70	EN80	EN90	FN00	FN10	FN20	FN30	FN40	FN50
EM69	EM79	EM89	EM99	FM09	FM19	FM29			
EM68	EM78	EM88	EM98	FM08	FM18	FM28			
EM67	EM77	EM87	EM97	FM07	FM17				
EM66	EM76	EM86	EM96	FM06	FM16				
EM65	EM75	EM85	EM95	FM05	FM15				

2) GRIDS HEARD OR WORKED ON TWO METERS BY WB2AMU (FN30)

EN65	EN75	EN85	EN95	FN05	FN15	FN25	FN35	FN45	FN55
EN64	EN74	EN84	EN94	FN04	FN14	FN24	FN34	FN44	FN54
EN63	EN73	EN83	EN93	FN03	FN13	FN23	FN33	FN43	FN53
EN62	EN72	EN82	EN92	FN02	FN12	FN22	FN32	FN42	FN52
EN61	EN71	EN81	EN91	FN01	FN11	FN21	FN31	FN41	FN51
EN60	EN70	EN80	EN90	FN00	FN10	FN20	FN30	FN40	FN50
EM69	EM79	EM89	EM99	FM09	FM19	FM29			
EM68	EM78	EM88	EM98	FM08	FM18	FM28			
EM67	EM77	EM87	EM97	FM07	FM17				
EM66	EM76	EM86	EM96	FM06	FM16				
EM65	EM75	EM85	EM95	FM05	FM15				

Grids heard or worked

Figure 4-3. The two plots above represent the grids that were heard or worked from WB2AMU's location in FN30 on 6 and 2 meters during the tremendous aurora opening of March 31, 2001. WB2AMU used moderate power levels and portable three-element Yagi beam setups on masts up 10 feet for both bands. The antennas were always pointed north, in the direction of the aurora. Aurora signals on 2 meters were particularly wide and distorted, so all contacts were made in CW.

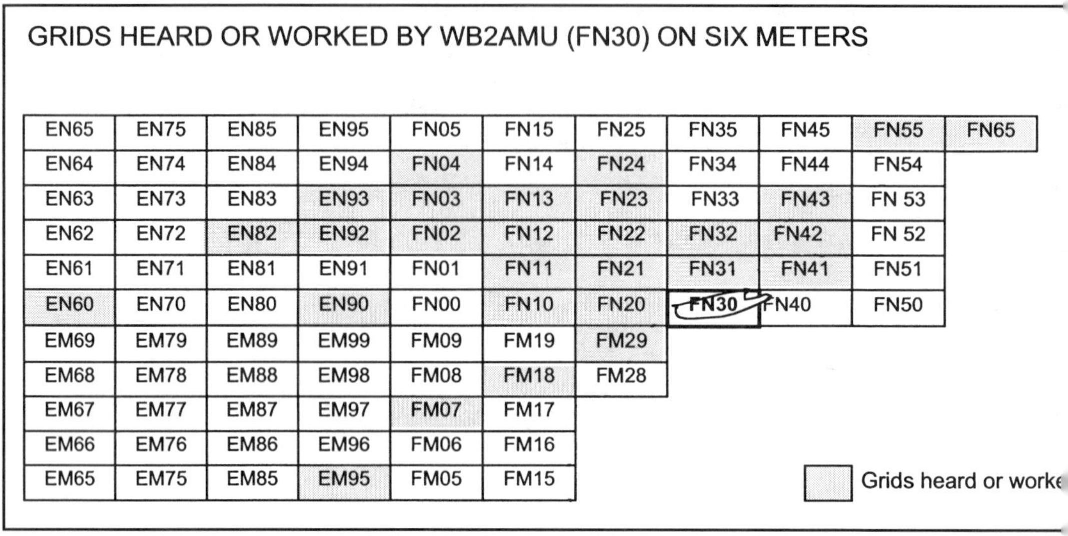

*Figure 4-4. The plot above represents the grid squares that could be heard or worked from WB2AMU's QTH in Long Island (FN30) during the April 6, 2000, opening, which lasted from 3:30 p.m. local time to 8:15 local time. Many of the grids worked during this opening were in "in-between" locations (those not reachable by line of sight and too close for sporadic-*E *skip). The opening was very intense with many signals present and the K index reaching 8. After the aurora opening faded, strong* F_2 *activity in much of the United States into Central and South America occurred for several hours, well past midnight, local time.*

2 meters, with three-element beams for both bands up at 10 feet. The opening occurred during a particularly active period (late March into early April) of that year and was unusual in that it occurred during the daylight hours. Not only was the opening very strong (it reached 144 MHz), but it was also of long duration, lasting over four hours! (See **Figure 4-3** for the grid squares that could either be worked or heard.) During this unusual period, there were actually reports of 432-MHz aurora contacts being made.

When aurora reaches the 2-meter band, it takes on even more importance! It becomes a viable means for working significant distances into areas that are not often heard on the band. During a strong aurora opening on April 11, 2001, for instance, Ken, WB2AMU, was able to work into Indiana on 2 meters with a portable setup. The direct path is very unlikely on 2 meters and so the distance covered—over 1,500 kilometers—was impressive.

What about hams located in the aurora zone? Where do they point their antennas? Judging from the experience of ardent VHFer Howard, WB4WXE, when he was stationed at North

Pole, Alaska, during the late 1970s, you should point your antenna in the direction of the signals heard. Essentially, they are inside the skip zone. Howard indicated that there were very few stations to work to the north of him and he just pointed where he heard signals. In his case, this was often east or southeast in the area of Minnesota, North Dakota, Alberta, Saskatchewan, British Columbia, or even Northwest Territory (VE8BY).

For hams located in either temperate zone, aurora activity generally occurs only around the equinoxes and may only occur two or three times a year during the low end of the sunspot cycle (hams inside the aurora zone see more activity than those in the temperate zone). As the sunspot count increases, more events are observed and can occur at times other than the equinox months of March and September. During the summer of 2000, a very high sunspot count year, many U.S. states in the northern part of the temperate zone experienced several 6-meter openings, with a major opening in the middle of July.

High power, on the order of 100 watts, is generally required for most openings because of distortion and signal fading. Some aurora openings can be quite strong, however, and may require power levels of only 10 watts. On a number of occasions during particularly strong openings, Ken, WB2AMU, worked stations in Maine and elsewhere from his location on Long Island using CW and a dipole antenna.

An additional side effect of a particularly intense aurora is that the *F* layer around the equatorial zones is energized by the same geomagnetic activity. So, as a rule of thumb, after a very strong aurora opening listen for F_2 activity running north and south. Ken observed an example of this during the April 6, 2000, aurora opening (illustrated in **Figure 4-4**). After this particularly strong opening, which reached 8 on the K scale, subsided at 8:00 p.m. local time, signals from South America started coming in for several hours. Ken worked HP3XUG from Panama and HK3YH from Colombia at 12:30 a.m. local time using only 10 watts and a dipole! (This phenomenon will be explained further in the chapter on F_2 propagation.)

Auroral-*E*

Another effect that can occur after an aurora opening is known as auroral *E*. This is when aurora signals become crystal clear with high-fidelity qualities. Such openings are more common in the higher latitudes, but some very long-range contacts can be made on 6-meters from Maine into British Columbia late at night via this mode. Six-meter operators have observed *E*s-type openings develop under these conditions and, by changing the direction of the beam, notice differences in signal quality.

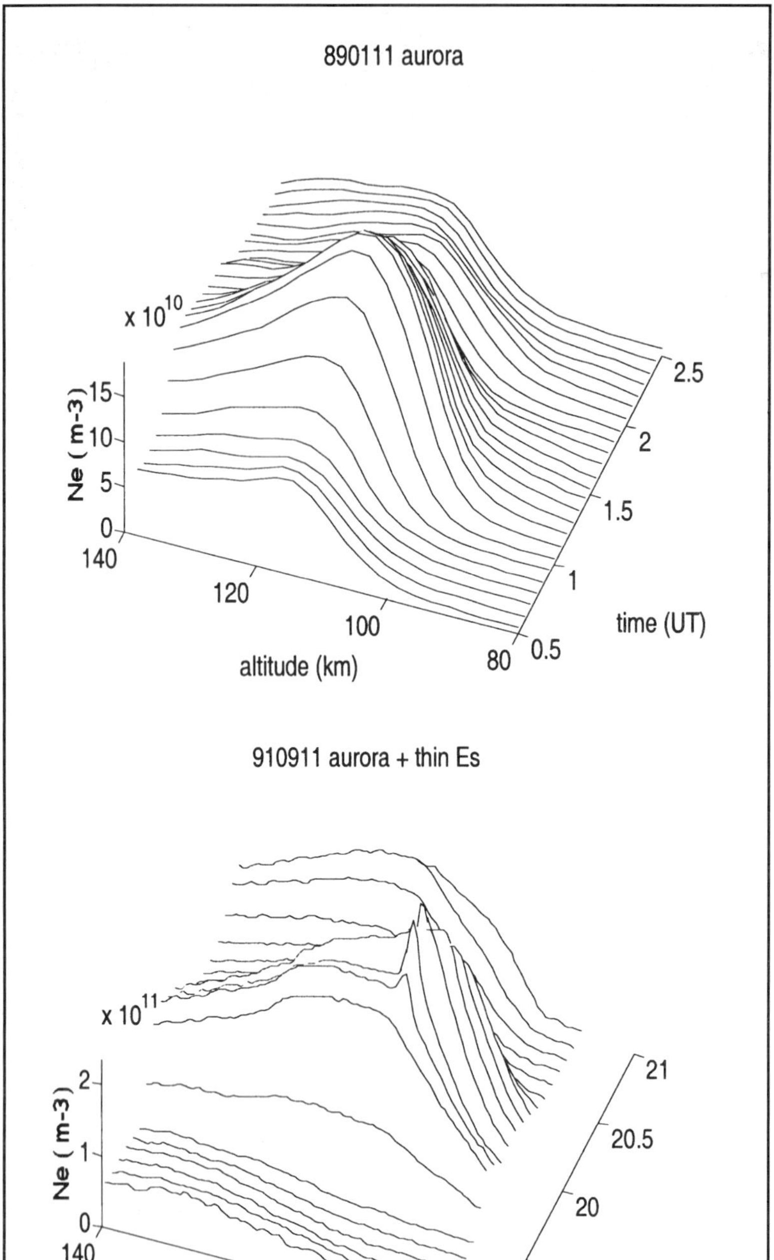

*Figure 4-5. Illustrations of the wide pattern of aurora. In the second diagram, note the thin sporadic-*E *formation below the aurora formation. (Figures courtesy of S. Kirkwood)*

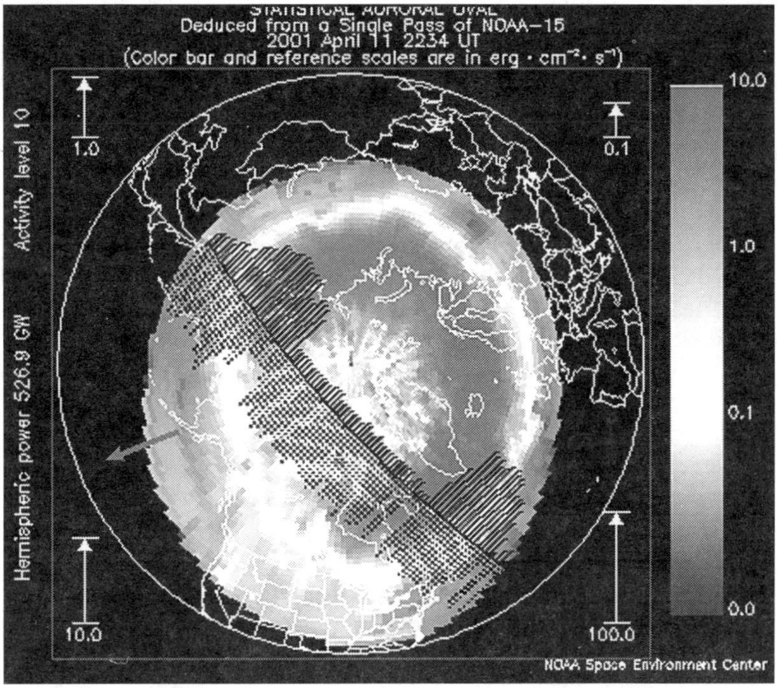

Figure 4-6. NOAA's NOAA-15 satellite flies over the earth's poles collecting data that help plot aurora behavior during periods of high geomagnetic activity. This particular plot shows very high aurora activity over much of northern Europe and Canada during the latter part of the day on April 11, 2001.

Just as with *E*s, aurora formations can be tracked by radar. **Figure 4-5** shows the pattern of the aurora over time as tracked by EISCAT radar. Notice that in comparison to the EISCAT radar figures of the thin *E*s formations in the previous chapter (**Figure 3-11**), aurora is a broader and seemingly less-focused formation. In the second plot of **Figure 4-5**, an *E*s opening can be seen in the lower part of the aurora. This plot appears to show an auroral-*E* opening and indicates that a process is taking place involving the metallic ions of the *E* region (where eventually the *E*s formation is dominated or absorbed by the aurora).

Is Aurora Predictable?

Potential aurora occurrences can be predicted a few days in advance by watching certain types of solar activity. By consulting Internet websites and monitoring NOAA data tracking (see **Figure 4-6**), it is possible to watch solar disturbances, such as solar flares and coronal mass eruptions (CME), as they happen. It generally takes two to three days for these events to travel the distance from the sun to the earth's ionosphere in the area above the poles, ultimately extending the aurora zones into lower latitudes.

An important parameter of aurora is the all-important Bz value. Measured by scientific instrumentation, this value refers to the direction of the magnetic field intensity in either a north or south direction, with direction determined by the sign of the value (a negative value indicates a southward direction). For stations in the Northern Hemisphere, it is best to have a Bz value favoring a sustained southward direction, which is conducive to producing geomagnetic storming southward from the geomagnetic North Pole.

For example, a major geomagnetic event took place on October 28, 2003, with the K value reaching 9 for several hours. However, because the Bz value was not negative for a period of time, it took several hours after solar impact on the magnetic field for major aurora events to be observed on 6 and 2 meters in the Northern Hemisphere. This is in contrast to one solar rotation later in the year, on November 20, when aurora events were observed very shortly after solar impact, because the Bz value was negative (south) even though the K value topped out at 8.

In Summary

Aurora is perhaps the strangest of all of the propagation modes that hams will encounter on any of the bands. It's fortunate that hams have 50 MHz and 144 MHz as part of their band allocations, as these are two of the best places to observe this mode and to gain a better understanding of its behavior.

References

13. L. Hesse, "VHF Bistatic Aurora Communications as a Function of Geomagnetic Activity and Magnetic Latitude," Artic Communications, Pergamon, 1964.

CHAPTER 5

METEOR SCATTER: SPACE ROCKS BURNING UP IN THE SKY

Meteor scatter (abbreviated Ms) propagation is another major VHF propagation mode that occurs in the *E* region of the ionosphere, much like the sporadic-*E* (*E*s) and aurora (Au) phenomena. This mode of propagation can appear on the amateur VHF bands beginning on 6 meters and going up as high as 432 MHz. It is an interesting mode in that it can allow for long-distance contacts on bands like 2 meters that cannot normally be accomplished by any other means.

Meteor activity occurs in the *E* region of the ionosphere when incoming meteors traveling from space enter at high speed and ablate, or burn up. When the meteor particles burn, they can cause long trails that can reflect radio waves that can last from one to several seconds each.

There are two types of meteor activity: sporadic and meteor shower. The sporadic type can occur at any time on any day, while the meteor shower type sees intense periods of activity that can occur only during certain times of the year.

There is not too much mystery regarding the actual composition of meteor particles as many specimens have fallen to earth and have been analyzed. More than 95 percent of all meteorites that fall to earth are of a stony composition. This can further be divided into two categories: chondrite and achondrite. Both types are composed mostly of silicate minerals, but the great majority of the specimens found also contain some traces of metallic iron and nickel.

Those larger iron-nickel specimens that do fall to earth and have been retrieved reveal a sort of geometric or crystal-like formation, known as Widmanstatten pattern, that is the result of heat and compression when the original meteor existed in space.

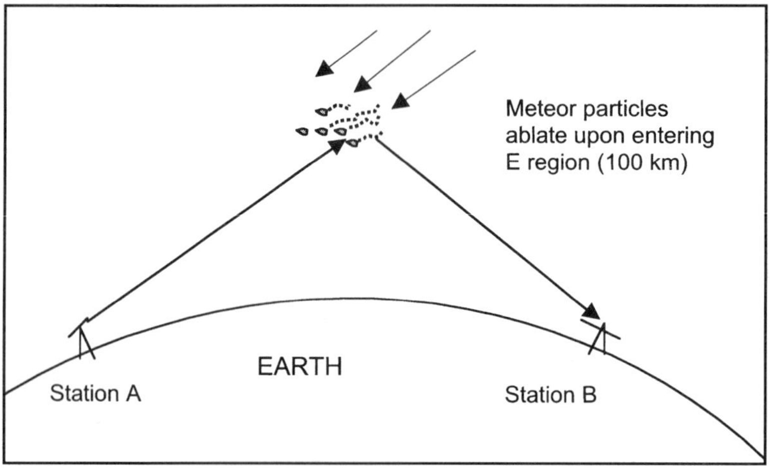

Figure 5-1. An illustration of how the signal from Station A is reflected off the trails of incoming meteor particles as they ablate (burn up) in the E *region of the ionosphere.*

In addition to these terrestrial samples, rocket launches have verified the presence of both nickel and iron in the composition of meteor particles when they are in the ionosphere (see **Table 3-1**). This information has been key in determining the origin of *E*s ions as there are similarities between meteor particles and *E*s ion composition. So, in a sense, these two modes of propagation (Ms and *E*s) are like first cousins.

When many particles are entering the *E* region, many reflections become possible, which may allow for the completion of a contact. The signal may be loud for only as long as five seconds, so contacts have to be made quickly. Sometimes during an intense meteor shower, the great number of meteor particles, and, therefore, reflections, can give signals the effect of constant rising and fading. It's like riding a wave when working other stations by this mode. The bursts of signals heard during meteor activity is generally referred to as "pings," and hams who are active in meteor scatter work are known as "ping jockeys."

VHF Propagation Audio Course available. See Appendix II for details.

Compared to *E*s and aurora, meteor scatter propagation is a relatively new mode to hams. The effects created by meteor burn were first heard on 6 meters in 1946, shortly after hams got on the band. By the fall of that year, several meteor contacts were made on six. It took a little bit longer, but by 1953, 2-meter meteor contacts were made as techniques and equipment improved. It would be several more years before 220-MHz and 432-MHz meteor contacts were made.

The greatest contribution in the propagation of an Ms reflection is the amount of ionization from the meteor entering our atmosphere when it burns up. When this happens, we can actually see a streak of light in the sky as the meteor mass gives off heat caused by air friction and the high temperature present in this region. Ultimately gases that have become ionized act as a short-lived reflector cloud within the ionospheric *E* layer.

Like aurora propagation, Ms is a visual phenomenon that can be observed in the sky under the right conditions as an interesting display, particularly during the major meteor showers. The following list shows shower activity during the course of the year that can be observed in the Northern Hemisphere. Approximate time periods are provided and major Ms events are indicated in bold:

Name	**Dates**
Quadrantids	**January 1–6**
April Lyrids	April 19–24
Eta Aquarids	**May 1–8**
June Lyrids	June 10–21
Ophiuchids	June 17–26
Capricornids	July 15–August 15
Iota Aquarids	July 15–August 25
Perseids	**July 25–August 18**
Chi Cygnids	August 19–22
Orionids	October 16–26
Taurids	October 20–November 30
Cepheids	November 7–11
Leonids	**November 14–19**
Geminids	**December 7–15**
Ursids	**December 17–24**

For many VHF hams in the Northern Hemisphere, the Perseids fall around August 12 and the Leonids fall around November 17 each year.

Setting up a Meteor-Scatter Contact

While random contacts are possible and have been made via meteor scatter, the vast majority is made by pre-arranged schedules between stations. These schedules are typically made by phone, e-mail, or Internet sites in anticipation of the major meteor showers. When a major meteor shower occurs, the rates of meteors entering the ionosphere are high enough for a contact to be completed. It may take several minutes and several transmissions for a contact to be completed on 6 meters. Typically, it takes longer for a contact to be completed on 2 meters than on 6 meters, because of the shorter burns that are experienced on that band.

As previously noted, sporadic meteors enter the ionosphere every day (on the order of billions of meteor impacts), thus schedules are also attempted at times other than the major showers using only the ambient daily meteor conditions. However, these sporadic meteor contacts may take much longer to complete, as there are fewer particles present to sustain a contact compared to the number present during a major Ms shower.

As there is a physical limit to this propagation mode, schedules should only be made with stations within a 1,500-mile radius. This is because Ms, like *E*s, is an *E*-region phenomenon, taking place at about 100 kilometers above the earth. As geometry shows us, that 1,500-mile distance is about the maximum that can be achieved (see **Figure 5-1**).

Can distances greater than 1,500 miles be covered by Ms propagation? The answer is probably no. That 100-kilometer height of the phenomenon sets the physical or geometric limits to the maximum distance that can be covered by a single hop. Greater distances would require conditions similar to double-hop *E*s; for instance, where Ms would have to also occur at another location to allow the second hop. While physically possible, this is highly improbable as it would require major meteor shower activity across a very wide area exceeding 2,000 miles. In the meantime, the single-hop range of Ms is particularly useful to VHF hams on 2 meters and above as new areas open up to be worked, compared to propagation modes such as tropo.

Meteor-Scatter Modes

So what is the best mode for meteor-scatter work? In theory, all modes of transmission, such as SSB, CW, and digital, can work under the right conditions, though the greatest success has been achieved with digital, particularly on the VHF bands above 6 meters.

For the SSB mode, a moderate meteor trail on 6 meters may allow for a contact that lasts up to 30 seconds. On 2 meters, the contact may last for only five seconds, and on 222 and 432 MHz any voice contact is so short that not much meaningful information may be exchanged. Thus for these higher bands, non-voice modes, such as high-speed CW (HSCW) and digital, will yield higher success rates in contact completions.

Meteor-scatter CW contacts at modest paddle and keyer speeds of 40 wpm will certainly have an edge over voice, but even better success is achieved with HSCW at speeds above 100 wpm. Some years ago, such contacts would have had to be tape-recorded and played later at half speed to verify the information exchange. However, with the advent of computer programs for ham radio, HSCW can be used in computer-to-computer contacts that are far more reliable, even when propagation conditions result in signals coming in just above the noise floor. The use of HSCW and other digital methods for Ms work is designated as HSMS, for high-speed meteor scatter.

New digital modes are also now being used for meteor contacts. One relatively new technique for meteor-scatter communications is WSJT, which stands for weak-signal communications developed by Joe Taylor. Joe (callsign K1JT) is a Nobel Prize science winner and has done much work in the field of VHF communications. WSJT uses frequency shift keying (FSK) of four specific tones. To work this mode, you would need a robust computer and a sound card interfaced to your computer. This setup allows for the transmission of this specific type of FSK and the translation of FSK signals received. Gordon, WB6NOA, has used the West Mountain Radio Rigblaster Plus sound card interface to achieve amazing results versus traditional voice contacts; however, he (and others) points out that the excitement of establishing an Ms contact on voice is far more thrilling than simply watching a computer knock off the QSO.

Conventions Used in Meteor-Scatter Contacts

A successfully completed contact is traditionally defined as one where the callsigns of each station are copied correctly and there is an exchange of some piece of information (such as signal report or grid square), followed by acknowledgement of copying the information, such as a roger (R) sign. While this may seem simple enough, it can actually take several minutes for this information to be successfully exchanged during an SSB Ms contact under marginal conditions!

One convention used specifically for meteor contacts is a unique signal report. Just like the letter A is used in aurora contacts, there is a specific letter and number used to give a signal report for a meteor contact. Remember, signals can go from very weak to a tremendous burst lasting a few seconds, so it's difficult to assign an average or true signal report.

For many years, an S report would be used for SSB and slow CW. The S report is based on a scale of S1 through S5, representing the burst length. The traditional S report that has been used for five decades by VHF hams is as follows:

S1 – ping with no information (not sent)
S2 – ping, up to 5 seconds in length
S3 – 5 to 15 seconds in length
S4 – 15 to 60 seconds in length
S5 – over 60 seconds in length

Thus, if you hear a station transmitting S2 over and over again during a CQ, you know that the station is hearing pings of up to five seconds long. During a major meteor shower where the rates are higher, it would not be unusual to hear a station using S3 or S4.

With the advent of HSMS modes, the signal report convention has been modified to a two-number code using information from the list above as the first number plus additional information in the following format:

First Number (Ping Duration)	**Second Number (Signal Strength)**
1 – ping, no information (not sent)	[no second number]
2 – ping, up to 5 seconds	6 – up to S3
3 – 5 to 15 seconds	7 – S4 to S5
4 – 15 to 60 seconds	8 – S6 to S7
5 – over 60 seconds	9 – S8 and stronger

Thus, one might copy a report of 27, which would indicate pings of up to 5 seconds at signal strength 4 to 5.

It is usually not recommended to use the calling frequency for scheduled Ms contact attempts. Typically, contact attempts on 6 meters would be in the 50.270- to 50.300-MHz range, and on 2 meters in the 144.100- to 144.140-MHz range.

There is also a convention regarding the sequence of transmissions for the different modes used in Ms contacts. This is that the western-most station would be the first to transmit for a specific period of time, followed by the eastern-most station's transmission during the next block of time. This specific period of time varies depending on which mode is used with the following examples typical:

First 30 seconds of each minute for FSK441
60 seconds for HSCW (with the western-most station calling on the first minute of each hour and half-hour).
15 seconds of each minute for SSB contacts
15, 30, or 60 seconds are used for slow CW

This points out the need for an accurate station clock based on atomic time when doing Ms work.

Other Considerations

It is usually necessary to run moderate to high power into directional antennas when chasing Ms contacts. This is because there is variable fading (QSB) on the meteor burst. This is not to say that QRP power Ms contacts cannot be made. Ken,

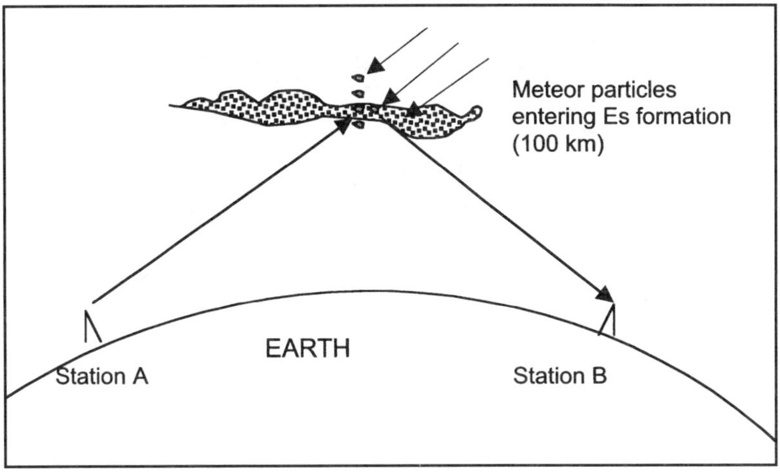

Figure 5-2. An illustration of how an Es *formation is enhanced by incoming meteor particles as they ablate (burn up) in the* E *region of the ionosphere.*

WB2AMU, has made occasional contacts while operating in the QRP category of the ARRL VHF contests held in January, June, and September when there are long bursts. Of course, these contacts have to be made quickly, usually in a matter of seconds, as signals fade rapidly.

Another important aspect of Ms propagation and the successful reception of a distant station is the angle of incidence for the signal of both stations in relation to the particles as they burn up in our ionosphere. Both stations must be in position to have a mutual "view" of the incoming meteor trails in order to establish successful two-way communications.

Perhaps the most important thing to remember about meteor showers is that their rates can vary from year to year, something scientists are working on trying to predict. Their rate strength depends on the orbital path of the meteor particles in the solar system and at what point they intersect with earth's orbit around the sun. The point of intersection determines whether the thickest part of the meteor group is encountered. At this intersection there is a greater density of particles so more enter the ionosphere and ablate. It's quite tricky to predict what part of earth will see the best action until the intersection actually occurs.

One last interesting aspect about meteor showers that we'll cover seems to be related to the summertime *E*s season. This condition occurs when the MUF of *E*s clouds that may be present in the area approaches 50 MHz, but does not reach it. However, ionization coming from an incoming meteor trail may induce some ionic action with *E*s ions, resulting in a brief *E*s-

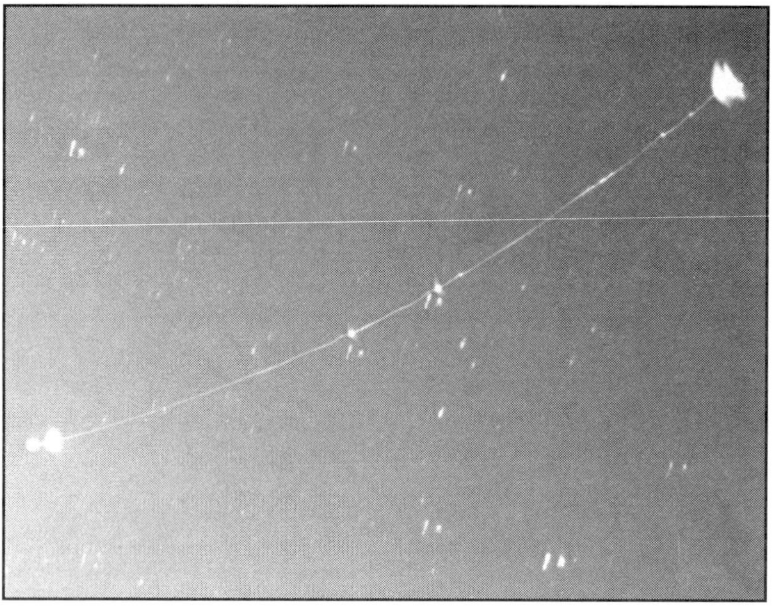

Photo 5-A. This is a shot of a long meteor burn taken by Gordon West, WB6NOA, from the California desert during the Leonids meteor shower in November of 2003.

type opening (see **Figure 5-2**). So far, this has been based primarily on amateur radio observations, and more scientific data is needed to confirm this condition.

Summary

Meteor scatter is the most challenging of all of the VHF propagation modes because of the short duration of the meteor burns (see **Photo 5-A**). It has always been painstaking to make these contacts via conventional modes, such as slow-speed CW and SSB. However, the initiative of hams using newer digital technology and new computer software has led to the development of more efficient methods of meteor communications, such as HSCW and WSJT. Indeed, the pursuit of Ms contacts has led to more technological advances in equipment and transmission modes than have scientific advances in the phenomenon itself.

CHAPTER 6

WAITING FOR THE SUN: F_2 PROPAGATION

It is through the F_2 propagation mode that some of the truly long-range DX can be worked on 6 meters. It's made for some extremely exciting activity during the peak of the sunspot cycle and has proven the worth of the 6-meter band.

Six-meter F_2 propagation was first discovered in 1946 when a cross-band contact was made between Ed Tilton, W1HDQ, who was transmitting on six, and G6DH, who was listening on six and transmitting on ten (UK stations were not allowed to transmit on 6 meters at that time because of the proximity of Channel 1). By the end of the decade, other F_2 paths were discovered on 6 meters. Since then, other countries have given 50-MHz privileges to their hams, leading to still more discoveries about F_2 on 6 meters!

F_2 and Ham Radio

Stations in the U.S. can work stations in Europe, Africa, South America, and even in areas of the Pacific with moderate equipment via this mode. In fact, many 6-meter operators wait patiently for the solar sunspot peak to arrive with its high sunspot count and solar flux values, just to take advantage of F_2. During the winter months of the sunspot peak, the higher solar activity energizes the oxygen molecules in the F_2 layer, located about 200 to 300 kilometers above the earth. The resultant high-electron density formations of these molecules allow radio waves up to certain frequencies to be reflected. The MUF represents the highest frequency that can be reflected.

Compared to the nuances of sporadic-*E*, the characteristics of 6-meter F_2 are more straightforward. F_2-based propagation follows the path of the sun, particularly on the HF bands. It takes the right conditions (generally high solar flux values along with high sunspot count, both exceeding values of 200) to create high electron densities in the F_2 layer. The increased electron densities make the MUF of the F_2 layer climb from the last HF amateur radio band of 28 MHz all the way to 50 MHz.

It's important to realize, however, that even terrific conditions on 10 meters don't necessarily translate into F_2 activity on six. Essentially 6 meters is just the tip of the F_2 propagation iceberg and is affected only a small percentage of the time compared to HF.

In contrast to *E*s signals, F_2 signals tend to be steadier with less fading while the opening lasts. Fading will occur as the opening gives out. Signals can reach as high as several dBs over S9, even with a mobile antenna, during optimal conditions!

VHF Propagation Audio Course available. See Appendix II for details.

It's not really easy to predict possible F_2 openings on 6 meters based on what's happening on 10 meters. That 22-MHz difference is significant, plus F_2 on 10 meters tends to be much wider in coverage. Most 6-meter operators will, therefore, use certain indicators to help them. For instance, they may monitor 40 to 50 MHz for known commercial signals, such as TV video signals from Europe, which are located between 47 and 48 MHz, or various state highway patrols that use 42 MHz. Even when these signals are heard, though, it does not necessarily mean that F_2 will occur on 6 meters.

As is the case of much HF propagation, 6-meter F_2 skip is primarily a wintertime phenomenon (often called the winter anomaly). The structure of the ionosphere during the winter is more conducive to F_2 skip at higher frequencies than it is during the summer. This is because the peak electron density of the F_2 layer is higher and occurs at lower altitudes in winter than in the summer, when F_2 peak electron density is fairly small and located at very high altitudes. Typically on a band like 10 meters, F_2 skip starts appearing in late September and lasts through late April. On 6 meters, the season starts later, usually in mid October and may last into early February. On occasion, some F_2 (usually north-south paths) may be seen during March and April at some locations in the world.

The distance that can be covered by a single-hop F_2 skip on 6 meters is typically on the order of 3,000 miles (see **Figure 6-1**). It's not uncommon, therefore, for eastern U.S. stations to work the West Coast or into western Europe with one F_2 skip. So when an F_2 opening occurs, it is a great opportunity to work a lot of new countries.

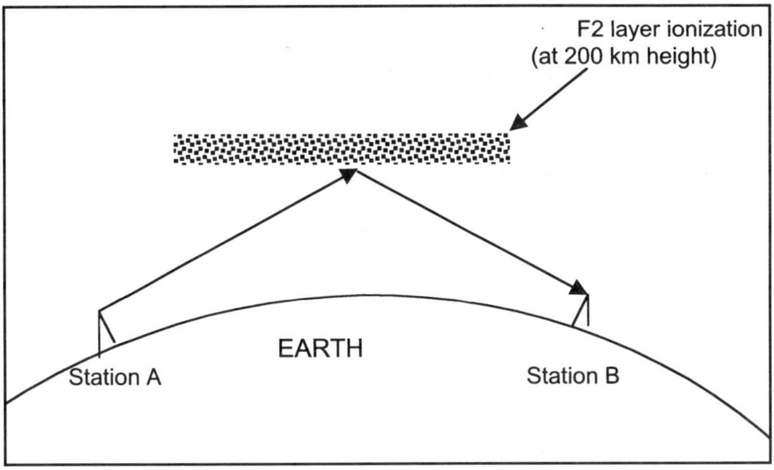

Figure 6-1. An illustration of how signals are reflected off the F_2 *layer of the ionosphere.*

Since it follows the path of the sun, F_2 on 6 meters is primarily a daytime phenomenon, with one major exception. In the event of a major solar flare that causes significant aurora activity on 6 meters, there is a good probability of F_2 propagation occurring in a north-south direction at some point after the activity subsides. This can happen at any time, even late at night.

The major aurora opening of April 6, 2000, which was discussed in the last chapter, illustrates this. Almost immediately after the aurora opening subsided (after giving hams about four hours of activity), stations from Central America and the northern part of South America were coming in by the dozens to the eastern United States, and at good signal strength. U.S. stations that had been pointing their beams north toward the aurora were now pointing south and a major pileup ensued. Ken, WB2AMU, was able to work HP3XUG in Panama and HK3YH in Colombia on CW using 10 watts and a dipole at 12:30 a.m.!

This particular type of F_2 propagation may be induced by the same high geomagnetic activity that causes aurora and was first discovered on 6 meters as early as 1949. The *F* layer near the geomagnetic equator is highly ionized under normal conditions, and the additional geomagnetic activity enables the MUF of this specific ionization to reach the 50-MHz level. It is interesting to observe that as soon as aurora activity subsides, the F_2 activity towards the south begins! This may indicate some sort of interaction between the *E* region in the higher latitudes and the *F* region in the equatorial zone, perhaps in the form of electric current or electron transport. So it pays to monitor the 6-meter band after an intense aurora opening in the hopes of picking up signals via F_2 propagation.

The garden variety F_2 activity occurring on 6 meters, however, appears during the daylight hours in the winter season of high sunspot count years. It has a seasonal pattern in terms of the paths that can be worked. When it first appears in the Northern Hemisphere during the latter part of October and into early November, it is generally in a north-south orientation. Stations in the United States can work mainly stations in the Caribbean, Central America, and the northern part of South America. Towards late November through December and early January, the east-west paths become the primary ones, although north-south paths can still occur. Thus stations in the eastern part of the United States can work parts of Europe during the morning hours, and later in the afternoon can work transcontinental paths into the western United States. West Coast stations work the East Coast during their local morning hours and later into the Pacific and Japan during the afternoon, local time. The general rule of thumb is that the sun should be at the midpoint at local noon between stations on the east-west path.

A Band for All Seasons? A Personal Highlight

Many 6-meter operators who got on "the Magic Band" in the mid-1990s had not yet experienced significant levels of F_2 propagation there. So they were really looking forward to the sunspot peak of Cycle 23, when the top sunspot count was expected to occur in late 2000. Yet, the year 2000 saw only moderate levels of F_2 conditions on 6 meters during November and December, and they were nothing compared to what was seen during the last solar cycle (Cycle 22), much to the disappointment of many.

There were a few days in November and December of 2000 where East Coast U.S. stations worked into Europe and into the West Coast, but there was no sustained period of activity lasting more than a few days. The solar flux values during November and December rarely reached 200 (the magic number for F_2 activity on six) for any extended period. By the spring of 2001, when many scientists felt that the peak had passed, there were low expectations for the coming fall months. Most propagation experts expected a continued decline in solar flux values.

Well, surprise! It turned out that the solar flux values were consistently high for the fall of 2001, above that magic level of 200, and the 6-meter band exploded with F_2 activity! The increase in these solar flux values during the fall of 2001 suggests that Cycle 23 had a double peak configuration. (One other difference between this cycle and the previous one was that there were many more 6-meter operators to enjoy it, now that it's a common band on the newer HF rigs.)

Activity started up big time on Sunday, November 4, when there was a very strong one-hour European opening that allowed Ken, WB2AMU, to work several stations in the UK and into Denmark and the Netherlands using only 20 watts and a dipole. A week later, he experienced a major opening into the West Coast of the United States, which lasted for several hours during the early afternoon. From his car he worked into California, Oregon, and Washington while running 60 watts into a mag-mount vertical. It was a surreal feeling to work such distant QTHs on 6 meters while going mobile!

Over the next several weeks, Ken and other stations in the northeast United States fell into a daily operating routine. There would be some F_2 skip to Europe during the midmorning hours, around 9 to 11 a.m. local time (1400 to 1600Z). Then the skip would swing around towards the west by noon and the northeast stations could work into Washington, Oregon, California, and British Columbia. On a few occasions, the skip would stretch out and Ken would hear Alaska. He was finally able to work his

Photo 6A. How great were conditions on 6 meters during November 2001 through January 2002? Well the above collection of QSL cards represents a small sample of the several hundred contacts that Ken, WB2AMU, made during this amazing time period. F_2 *activity was a regular occurrence on 6 meters every day!*

first Alaska station, KL7FZ, on six, on November 12 at 1:30 p.m., local time. There were even reports that many West Coast U.S. stations were working into Japan later in the day via F_2 (see **Photo 6-A**).

During November and December of 2001, many U.S. stations heard openings into Europe that featured strong signals and lasted for two hours or more! Ken worked many stations using a vertical antenna while mobile in Long Island. Some days were better towards Europe in the morning, while some days were better towards the West Coast in the afternoon. It seemed like there was a consistent pipeline between East Coast stations and grid square CN87 in the state of Washington! Many operators were truly surprised at the consistency of the openings.

No doubt the Internet and the presence of so many more hams were key factors in the number of stations worked via F_2 compared to the previous cycle. It was amazing to observe F_2 propagation on a daily basis on 6 meters, an idea that would have been scoffed at by many in the past. It became a consistent band for long-range DX during this special time! As many as 30 days of F_2 were experienced during December, 2001, and many operators saw their DX country count on 6 meters starting to climb.

Some stations located in western Europe did exceptionally well on 6 meters during that season. For instance, Johan, ON4ANT, and Calum, GMØEMX, made over 1,000 QSOs with U.S. stations on 6 meters, working the majority of the FN and EM grid fields that represent much of the eastern United States! Many UK stations, including Neil, GØJHC, Conrad, GØRUZ, and Dave, MMØAMW, were on consistently during this period as well. After a while the pileups of early November subsided, and it became reasonably possible for any ham in the eastern United States with just about any kind of antenna to work some of these 6-meter signals from Europe.

How to Work F_2

The "double peak" configuration that occurred during Cycle 23 indicates that the months surrounding the predicted sunspot peak may not necessarily be the optimal time for 6-meter F_2. Indeed, it would appear that the next winter season following the peak (as seen in Cycle 23) might actually be best.

As we've seen, you don't necessarily need high power or super antennas to work F_2. During the fall of 2000, many stations were able to work stations in South America and Europe using simple antennas, such as mag-mount verticals mounted on cars or dipoles strung in a tree, along with moderate power levels from 20 to 50 watts. Weaker F_2 openings generally demand a

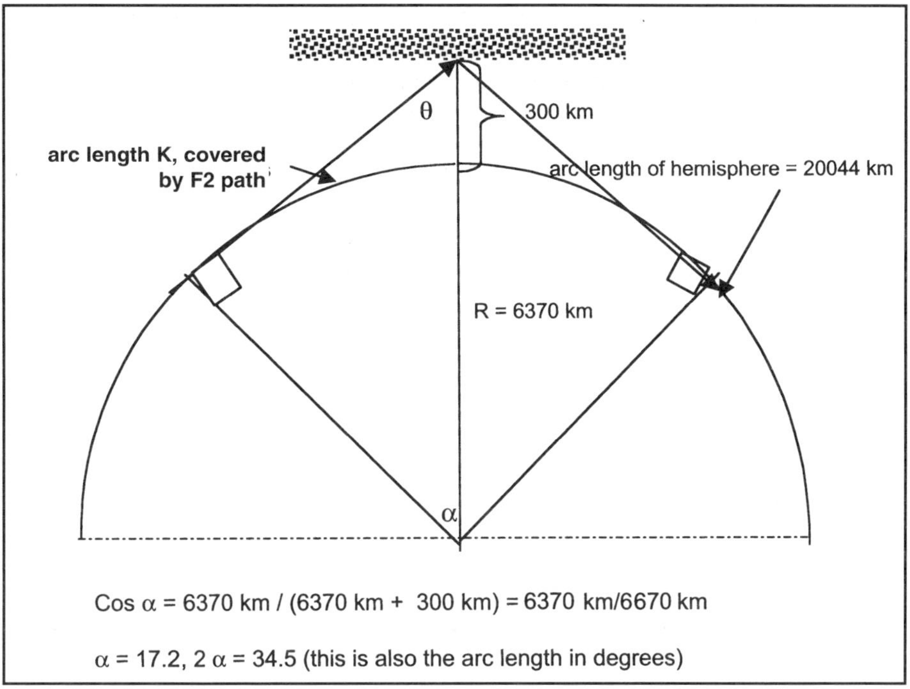

Figure 6-2. The above figure shows the maximum distance that can be covered by a single-hop F_2 *path. The calculations shown are similar to those performed for Figure 3-2. Using the same methodology used in Figure 3-2, an angle of 34.5 degrees yields a distance covered of approximately 3840 km.*

better setup for contacts to be made, and they'll also help break through pileups.

Following are some basic rules that apply to F_2 openings:

- Consecutive days of F_2 activity (in particular, those following east-west paths) generally require high solar flux values exceeding 200
- F_2 is primarily a daytime phenomenon, occurring from about 8 a.m. local time until 5 p.m. local time, with direction of the skip changing with the path of the sun (the one exception to this is an aurora-induced F_2 opening, which can occur at anytime)
- The peak of the F_2 season is generally December for stations in the Northern Hemisphere during the peak of the sunspot cycle. During high solar activity, F_2 can be observed daily on 6 meters during December, as was the case in 2001.

Backscatter is possible during a strong F_2 opening on 6 meters, just as it is on the HF bands. In general, the signals propagated by backscatter are weaker but can cover significant differences. For example, Mick, W1JJ, from Rhode Island, has heard stations as far south as Florida via F_2 backscatter while pointing east towards Europe. **Figure 6-3** illustrates F_2 backscatter.

During the major F_2 openings that occurred from November 2001 through early February 2002, the distances covered by F_2 skip were pretty consistent, typically ranging from the northeast United States into the UK and vicinity and from the northeast United States into the northwest. Very rarely did the skip distance shorten from this particular path. The reason is probably that the MUF was generally around 50 MHz and did not go much above that (the higher the MUF above 50 MHz, the shorter the distance of the F_2 path on 6 meters).

The geometry of F_2-layer propagation is similar to that of *E*s regarding MUF and critical frequency (f_0). The difference is that F_2 is situated at 250 kilometers instead of 100. Another major difference involves the behavior of F_2 itself: the MUF can reach 50 MHz (6 meters) but can't go much higher, as it can in *E*s. In fact, *E*s propagation can have an MUF starting around 21 MHz and can go as high as 220 MHz on rare occasions. That's why F_2 is generally only observed on one VHF band—6 meters—and covers very specific distances with minimal variations.

By traveling to different locations in the United States during the peak of the F_2 season, you can notice this specific distance aspect at work. For example, when Ken, WB2AMU, went to Nevada in November 2001, he could work or hear stations in eastern Canada and Maine but nothing further west of these locations. Many stations in the central United States could not work cross-country to either the east or west at all during the peak of the F_2 season. Specifically, a QSO between New York and Colorado or between Connecticut and Wyoming on 6 meters was generally a physical impossibility via F_2 propagation, because the MUF did not go much above 50 MHz. So, except for north-south F_2 paths, stations in the central United States may find occasional east-west contacts that may involve another mode of propagation, such as *E*s.

As the height of the formations associated with F_2 and *E*s propagation are different, so too are their geometric models. Instead of the approximation for the relationship between the MUF, or f_{max}, and the critical frequency, f_0, (vertical) for *E*s where:

$$f_{max} = 5.33 \times f_0$$

the equation for F_2 propagation is approximately

$$f_{max} = 3.4 \times f_0$$

Thus in order for the MUF to reach 50 MHz, the critical frequency (as measured vertically by an ionosonde) has to be in the area of 15 MHz or more.

While there is a well-defined relationship with frequency and distance between 6 meters and 10 meters in terms of *E*s propagation (as was shown **Figure 3-5**), the relationship between these bands is not as clearly defined with regards to F_2 propagation. The "shortening" effect that is seen in terms of distance on 10 meters during an *E*s opening, and which generally means that *E*s will appear on 6 meters, is not as clear with F_2 propagation. This is primarily because of its different geometry, where the higher altitude of F_2 formations allows longer paths to occur on 28 MHz, even when the MUF reaches 50 MHz.

This was illustrated when Ken, WB2AMU, participated in the ARRL 10-Meter Contest in December, 2001, and was simultaneously listening on 6 meters. When he detected an opening on six, where he worked KD6B in Oregon at 1 p.m. local time, he was able to work stations five minutes later on 10 meters in Utah, Arizona, Idaho, and Colorado. This shows the "shortening" effect on 10 meters, and certainly the direction of the F_2 skip. It's difficult to make too definitive a conclusion, however, as skip was widespread to other areas on 10 meters at the same time. Indeed, while operating 10 meters during the CQ World Wide DX CW contest in November, 2000, Ken worked HC8N towards the south on 6 meters, in between QSOs on 10 meters with European stations. At the same time, there was no indication of any southern paths on ten. So tracking F_2 on 10 meters is not that reliable an indicator for potential F_2 activity on six.

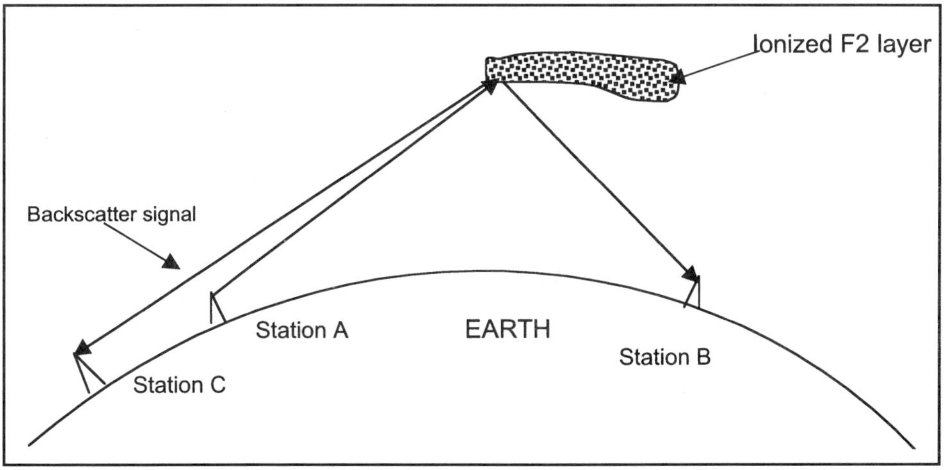

Figure 6-3. Station A is able to work station B by normal reflection off the F_2 *layer, while at the same time is able to hear Station C in the opposite direction via the backscatter mode.*

When solar flux values exceed 200 each day during June and July, you might wonder whether Southern Hemisphere stations, say in Argentina or Australia, experience F_2 east-west paths on 6 meters. Actually, there were only a few reports of any east-west F_2 paths occurring in this hemisphere during Cycle 23. The question is why.

The answer lies in two of the Southern Hemisphere's characteristics: 1) the geographic makeup of its land masses; and 2) the actual physical location of many of its 6-meter hams. A globe is helpful in understanding the possibilities for east-west paths in the Southern Hemisphere. To find substantial 6-meter ham representation in the south, you have to look primarily to the major population centers in Argentina, the East and West Coasts of Australia, and South Africa. Starting from the population areas of Buenos Aires, Argentina, and Montevideo, Uruguay, going towards the east, you'll find only a few locations in Africa that fall within one F_2 skip distance on 6 meters, and there are few hams in these countries.

In fact, for this population center in South America to work into South Africa, a center of some 6-meter activity, you would need an F_2 plus an *E*s link. This is would be highly improbable on the South African side as *E*s is very rare throughout the year at this location. Toward the west, 6-meter hams in South America have only a few islands in the Pacific to work.

For 6-meter stations in Australia, one possible F_2 east-west path is from New Zealand into western Australia. There aren't a lot of east-west paths from eastern Australia that cross major concentrations of 6-meter hams. A similar situation applies to South Africa. Thus much of the F_2 activity for stations located in the Southern Hemisphere is of the north-south or skewed variety.

It should also be noted that high levels of geomagnetic activity in the form of solar flares or coronal mass ejections (CMEs) will have an initial detrimental effect on VHF propagation modes, such as F_2, as it does on the HF bands. There may be periods of high solar noise on the bands. Listening to 10 meters after the effects of a solar flare is like listening to a quiet 6-meter band. During this time, there may be periods of auroral activity (based in the *E* region) developing on six.

F_2 propagation on 6 meters can also be wiped out by a solar flare. However, since the peak of the 6-meter F_2 season is usually from November to January during the peak sunspot years, this does not usually coincide with the peak time of solar flare activity, which primarily occurs around the equinox. (Observations have shown that 10-meter F_2 activity during the equinox period is often adversely affected by solar flares.) For the few times that high levels of geomagnetic activity occurs during the peak months of 6-meter F_2, the F_2 activity also picks up within a day or so, usually with north-south F_2 paths developing.

So when is the F_2 cycle officially finished on 6 meters? It's an interesting question. The answer appears to be when geomagnetic activity is significantly diminished. During the latter part of May, 2003, the solar flux values were less than 130, but there was some major earth-directed solar flare and CME activity. After the ensuing aurora resurgence, there were periods of F_2 activity and stations in the northern United States could work stations from Colombia and Venezuela.

In Summary

F_2 propagation is the mode that many 6-meter operators wait for with great anticipation during the peak of the sunspot cycle. As the solar flux values climb, operators check for potential signals developing in the 40- to 48-MHz frequency range. Some really good DX can be worked using the F_2 mode, but it appears only during a five-year window of the solar cycle, compared to the year-in consistency of *E*s. But when a good F_2 opening occurs, the opportunity to work new countries and grid squares is even better than on the *E*s mode!

CHAPTER 7

TRANSEQUATORIAL PROPAGATION

Transequatorial propagation (abbreviated TEP) is a unique form of *F*-layer propagation that involves the ionization of the *F* layer over the geomagnetic equator of the earth. It is primarily a north-south path and can cover some great distances on the 6-meter band. TEP typically occurs during the peak years of the solar sunspot cycle where it has been observed with regularity in certain areas of the world on 6 meters during the fall and spring seasons. TEP has also been observed on certain occasions and in certain areas on 2 meters, and it has even been known to reach as high as 432 MHz!

The earliest reported instance of 50-MHz TEP took place in October of 1947 when radio amateurs detected a number of different north-south paths. One path saw XE1KE in Mexico working LU6DO in Argentina, while in another area of the world many European stations worked ZS1P and ZS1T on 50 MHz. Hams continued to work TEP through subsequent sunspot peaks, and during Cycle 21, it made an appearance at frequencies higher than 50 MHz! Following are some examples of long-distance contacts on these high frequencies:

KP4EOR in Puerto Rico to LU5DJZ in Argentina on 145.1 MHz on February 12, 1978, at 0012Z, a distance of 6,340 km (new world distance record on 144 MHz)

YV5ZZ in Venezuela to LU3AAT in Argentina on 432.1 MHz on February 13, 1978, at 0110Z, a distance of 5,100 km (long distance 432-MHz signal reception)

5B4WR in Cyprus to ZE2JV in Rhodesia 144.1 MHz on April 10, 1978, at 1740Z, a distance of 5,800 km (first known TEP contact between Europe and Africa)

SV1DH in Greece to ZS6DN in South Africa on 144.1 MHz on February 13, 1979, at 1815Z, a distance of 7,120 km (new world distance record on 144 MHz)

It can be seen from the above examples and from the discussions in previous chapters that TEP is the only terrestrial propagation mode that can carry 2-meter signals beyond 1,500 kilometers (the range of a typical single-hop *E*s or Ms path). Indeed, it appears that TEP is the only form of *F*-layer propagation possible for 2 meters. It should also be noted that the paths for super DX on 2 meters are only north-south.

When TEP appears on 6 meters, it is an efficient form of propagation in that great distances—as much 6,000 miles—can be covered using moderate power levels. Like F_2 propagation, it has an associated Maximum Usable Frequency (MUF) and it can be observed from as low as 15 meters to as high as 2 meters, as we've seen.

Although TEP is a form of *F*-layer propagation, it is being treated in a separate chapter for discussion in this book. This is because TEP is not a conventional reflection mode like F_2; rather it is mode that allows the transport of radio waves over extended distances in the *F* layer without the use of ground reflection in between. In addition, TEP continues to appear on 6 meters for some years after the solar sunspot peak, long after most conventional F_2 propagation is gone.

The Mechanics of TEP

So how does TEP work in comparison to conventional F_2 propagation? As in the case of F_2, TEP takes place in the *F* region of the ionosphere at about 250 kilometers above the earth. And like F_2, it involves the ionization of oxygen ions in this region that have the capability of reflecting radio waves. But that is where the similarities between TEP and F_2 propagation end and the differences begin.

First off, TEP takes place in the equatorial zone of the earth, which is the area that follows the geographic equator. Then, instead of a radio wave simply being reflected off the *F* layer, TEP has a north and south entry point (located about 20 degrees from the geomagnetic equator) where radio signals that enter are tilted. When signals enter at these points, the signal is actually "trapped" or extended over the equatorial zone and can come down several hundred miles distant. At no point does the signal return to earth once it is in this zone (see **Figure 7-1**). This mode is also known as chordal hop, a reference to the particular tilt of the entrance and exit angles that do not allow the signals to return to earth.

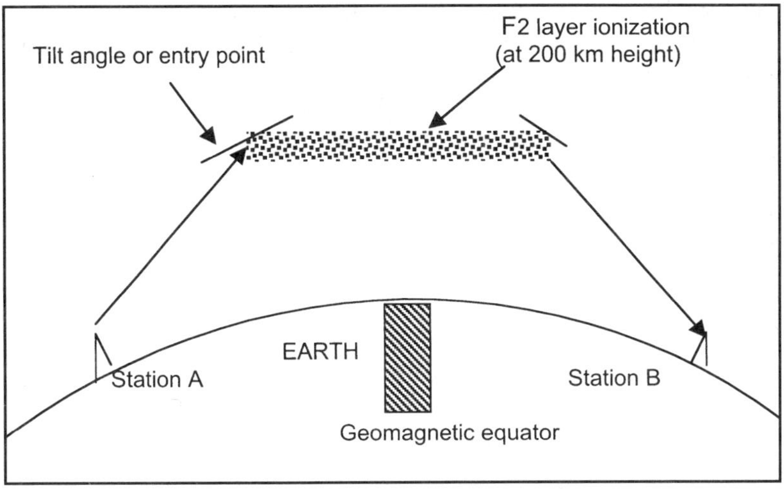

Figure 7-1. The figure above shows how the signal enters the F *layer of the ionosphere at a tilt angle and is extended over the geomagnetic equator to create a TEP path. The stations involved are located at equal distances from the geomagnetic equator and the overall distance is on the order of 6,000 miles.*

The area over the geomagnetic equator has been observed to have very high electron densities in the *F* layer, particularly during the equinox season, when geomagnetic activity tends to be higher than at other times of the year. Therefore, the best time for signals traveling this north-south path is generally during the months around the fall and spring equinoxes. The equinoxes are the best time for TEP-type skip on 6 meters because the winter anomaly effects of F_2 propagation overlap a bit in the Northern and Southern Hemispheres. At other times, when it is winter in one hemisphere and F_2 is possible (during the higher sunspot count years), it is summer in the other hemisphere and no F_2 is present; thus there is no chance for north-south F_2 paths between the hemispheres.

In many ways, the high level of ionization of the *F* layer over the geomagnetic equator resembles the high amount of electrical energy within the two aurora zones. The *F* layer in this zone has a sort of residual energy that can result in ionization that allows for radio propagation. It should be noted that TEP is a common mode of propagation on HF bands, such as 10 and 15 meters, throughout much of the solar cycle. Thus, this equatorial *F* layer seems to be energized even further during the high periods of geomagnetic activity that typically occur around the equinoxes.

As mentioned, TEP can allow for long-distance communications of over 6,000 miles between stations located equidistant from either the north or south entry point. Remember that we're talking about the geomagnetic equator and not the geographic equator. This fact is important because the geomagnetic equator varies significantly from the geographic equator as we circle the globe, particularly in the area of the Americas.

TEP usually has a unique tonal quality to it, with a touch of light distortion on the signal as a result of being extended through the *F* layer over the geomagnetic equator. This is not unlike the distortion effects encountered with aurora signals in the aurora zones, although aurora signals are significantly more distorted. TEP will generally be heard no earlier than noon local time (for the locations in the north-south path involved), usually appearing during late afternoons and continuing through the early evening. On some occasions, it can appear late into the night during a major opening.

TEP is almost a regular occurrence on 6 meters during the months surrounding the spring and fall equinoxes during the peak years of the sunspot cycle for stations in the TEP zone. It tapers off significantly towards the quieter years. Following are some of the common TEP paths observed on 6 meters worldwide:

Spain and Mediterranean area (40° N) to South Africa (25° S)
South Florida, South Texas, and Caribbean area (25° N) to Argentina (35° S)
Southern California (32° N) to New Zealand (38° S)
Japan (30° N) to Australia (30° S)

As you can see from this list, the geographic latitudes are different for TEP paths depending on location in the world. This is because, as previously mentioned, the geomagnetic equator does not follow a straight line around the world and can vary in location for different regions. For example, while the geomagnetic equator is closer to the geographic equator for the Europe to South Africa path, it is very different for North and South America. Here, the typical TEP path is from the area of south Florida and the Caribbean at 20° North down to Argentina at 35° South. The geomagnetic equator in this part of the world dips quite a bit (about 10 degrees from the geographic equator), because it is affected by the positioning of the geomagnetic North Pole, located about 15° North in this path (see **Figure 7-2**).

Because of this alignment of the TEP zones, it's possible for much of Western Europe to experience TEP on a regular basis on 6 meters, while only the southern portion of the United States can. Southern U.S. stations, such as Tom, K4SUS, in Miami and John, W5UWB, in southern Texas, do experience the TEP path to Argentina regularly on 6 meters, particularly during the equinox months of the higher sunspot count years. Stations in the southern part of California, below Los Angeles, such as N6XQ, not only occasionally observe paths into Argentina during the early part of the afternoon of the equinox months of these years, but also into the New Zealand and Australia area in the later afternoon and evening hours.

An interesting question, particularly for U.S. hams, is where does the "sweet spot" for the TEP zone end? Looking at Florida, we see that the far south area where Miami (26° North) is located seems to be in the active zone of TEP. However, we can go north another 200 miles and still hear almost regular occurrences of TEP in the Orlando area (28° North) on 6 meters. We can go even further north to the Jacksonville area (30° North) and still be able to experience TEP on an occasional basis. TEP on 6 meters has even been reported from southern Georgia on rare occasions. It appears then that the end of the TEP zone varies somewhat through the stages of the sunspot cycle, by as much as a 5-degree difference in latitude.

Working TEP

Moderate power is generally the rule for TEP contacts, though lower power contacts can also be made when conditions are good. For example, LU5VV in Argentina has made many con-

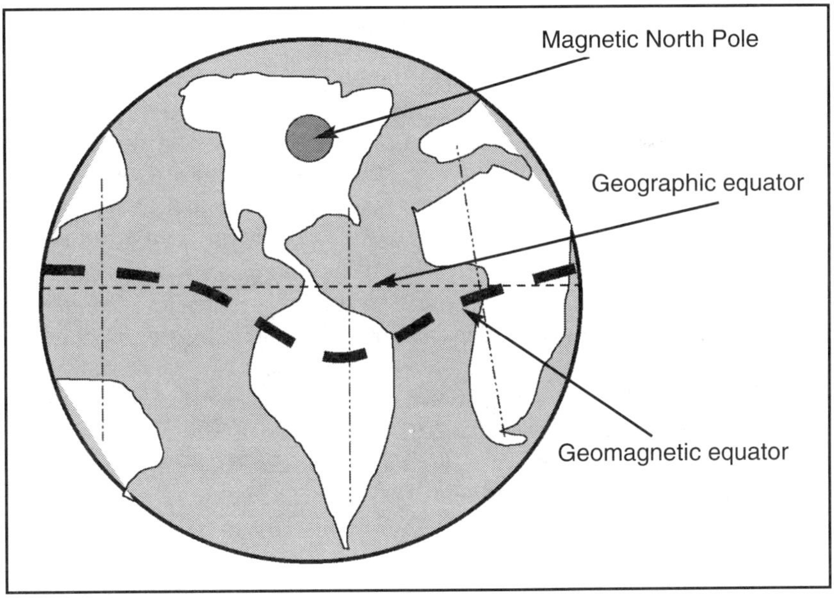

Figure 7-2. In the figure above some common TEP paths are depicted: Japan to Australia, Florida to Argentina, and Spain to South Africa. Note that the Florida to Argentina path is lower than the other two paths because of the dip in the geomagnetic equator.

tacts into the United States using only 5 watts and a five-element Yagi. Also, Jack, N6XQ, made many contacts into Japan via TEP when he visited Australia in 2001 using the DX handy (a vintage 6-meter QRP transceiver) with 1-watt output and a whip! Higher power is helpful, however, during marginal conditions when rapid fading is occurring.

This is further emphasized when low-power beacons can be heard via TEP. For example, when WB2AMU was visiting in the Orlando, Florida, area in November, 2003, he was amazed by how loud the LU1DMA beacon from Argentina was coming in at 4 p.m. local time. One reason for such efficiency of TEP signals is that the signal does not return to ground in between stations during the path. Any return to ground, as in multi-skip propagation modes, could result in some signal loss.

Another observation concerning TEP paths, such as that observed between Florida and Argentina, is that the solar flux values do not have to be exceptionally high, as they must be for east-west oriented F_2 paths. Indeed, TEP contacts have been made with the solar flux value just above 100, compared to solar flux values of 200 that are typically needed for the east-west F_2 paths. It should be noted that well after the major F_2 openings on 6 meters from Cycle 23 faded by the end of 2002, TEP propagation continued to appear occasionally on 6 meters into early 2004. Observations like these show how extensive the geomagnetic activity can be in the area of the *F* layer over the equatorial zones surrounding the geomagnetic equator, even during less active years of the sunspot cycle.

Some of the more intense TEP openings have occurred within 24 hours of a major aurora opening. The same geomagnetic phenomenon that causes intense aurora activity will often cause additional ionization in the area of the geomagnetic equator. So just like F_2 openings that have been induced in this way (as described in **Chapter 6**), TEP openings can occur with additional ionization of the *F* layer over the area around the equator. Many stations in the higher latitudes in the temperate zone in North America can look for this situation. In **Chapter 8**, we'll discuss how these higher-latitude U.S. stations can work stations in South America via TEP (as we'll see, it involves an *E*s link).

A paper by Jim Kennedy, K6MIO/KH6, titled, "50 MHz F_2 Propagation Mechanisms"[14] (a recent reprint of which also appears in the *CQ VHF* Winter 2003 issue) discusses the chordal-hop mechanisms involved in TEP. John also discusses some variations of TEP when, in addition to conventional north-south paths, some side scatter (where signals are directed at a sideways angle instead of directly) paths are possible for stations located in or near the TEP zone. For example, a station in Argentina may be able to work a station in Hawaii, or even Japan, on 6 meters by means of side scatter. Basically, the signal can be viewed as trapped, or retained, in the TEP zone where it travels along the equator until it exits at some point. At times, there may be some heavy modulation on the signals and they may sound distorted, like aurora or some Doppler shift occurrences.

For instance, when Ken, WB2AMU, was visiting St. Thomas in the U.S. Virgin Islands in December, 2002, he was treated to hearing the CW signal of ZD8Z coming in via TEP on 6 meters from Ascension Island, located south of the equator in the south Atlantic Ocean. ZD8Z's signal was very strong but had a slight wavering quality to it, almost like a tropo signal, giving it an otherworldly effect.

In Summary

We have seen that TEP propagation is the basis for some very long-distance communications on 6 meters and sometimes on two. We've also seen that it's "zone sensitive" and can only be experienced if you are located in the proper zone or are linked to it by another propagation mode, such as *E*s. In the next chapter, we'll discuss these combination modes along with long-path propagation.

It is important to note that any very long-distance propagation on 6 meters, where crossing of the equator is involved, will usually involve TEP or a very similar type of *F*-layer propagation (where the signal is extended in the *F* layer over the geomagnetic equator). Contacts made between the southern United States into Australia on 6 meters may involve a combination of F_2 and TEP, even though the TEP path is skewed somewhat. More observations will be necessary to better define all cases of TEP and similar paths that cross the equator on 6 meters.

References

14. Jim Kennedy, K6MIO/KH6, "50 MHz F_2 Propagation Mechanisms," *Proceedings of the 34th Conference of the Central States VHF Society*, July 2000.

CHAPTER 8

COMBINATION PROPAGATION MODES: MIX AND MATCH EVENTS

Some of the more interesting conditions on the VHF bands occur when two modes of propagation are linked together to allow long-range contacts. For the most part, combination events occur on the 6-meter band; however on rare occasions, they do occur on 2 meters also.

Just about any combination of propagation modes you can think of can occur at one time or another on 6 meters, but only under certain circumstances. The truly long-distance DX that's occasionally possible on six results from of a combination of propagation modes, sometimes referred to as "mix and match" propagation and designated with a "+" sign. This combination can only occur at certain times of the year and is most likely during the peak years of the sunspot cycle.

The best months for mix and match propagation modes (and consequently long-range DX on 6 meters) seem to be October, November, April, and May. This is apparently because certain modes of propagation are present during this time of year and have the potential to appear at the same time on any given day. Sporadic-*E* (*E*s), aurora (Au), F_2, and TEP are all possible during these months.

A case in point occurred when Ken, WB2AMU, was able was able to make contacts from his New York QTH with Argentina on a few occasions during these four months because of the TEP+*E*s combination. His most recent experience occurred on October 6, 2002, when he worked two stations in Argentina and PY5CC in Brazil. This particular opening was very strong and lasted for over two hours, and the *E*s opening lasted even longer, with QSOs going into many southern states.

The night before, aurora was observed, so the high geomagnetic activity associated with the aurora would also cause the *F* layer

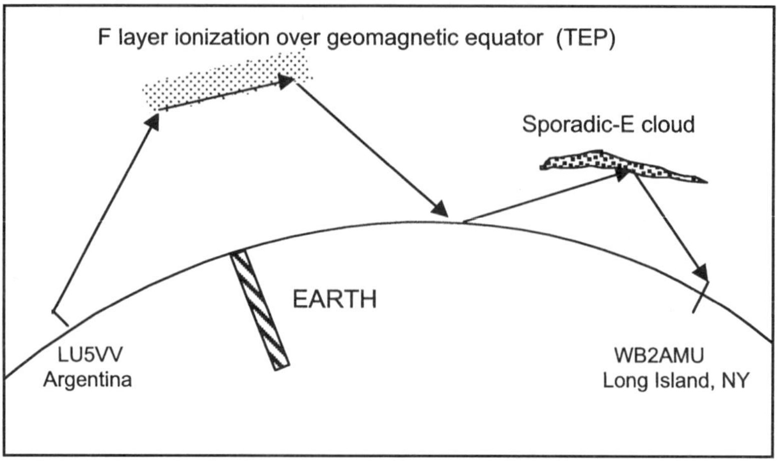

Figure 8-1. The above opening took place on May 25, 2000, at 6 p.m. local time. WB2AMU was running 50 watts from his car using a mag-mount vertical on the roof, while LU5VV used 5 watts into a Yagi.

over the geomagnetic equator to see an increase in ionization shortly after the Au opening. In this case, the TEP occurred about 20 hours after aurora was observed and, by chance, an *E*s link was present down to Florida at the same time (see **Figure 8-1**).

Ken worked another possible TEP+*E*s combination during a 6-meter DXpedition to Easter Island in March and April of 2001. On April 3, Arliss, CEØY/W7XU, was working into Florida at 1500 when an *E*s opening appeared and linked signals further north. First a station in Tennessee came in, then the skip settled in the New York, New Jersey, and Pennsylvania areas. Ken was on his lunch hour at his work QTH in Long Island and quickly put together his portable 6-meter setup: a two-element beam up 15 feet using five-foot pole sections and umbrella stand. He was surprised to actually hear Arliss on SSB and was able to contact him by CW. That this was an *E*s link was verified later when Ken worked AC4TO in Florida via a weak *E*s opening shortly after the Easter Island QSO and also heard other stations in Florida at the same time.

While Arliss was down on Easter Island in April 2001, he had several interesting experiences working a number of stations in India via long path on 6 meters. He would point his beam not towards India to the west, but towards the northeast. Although a number of the Indian stations used QRP power levels, they received decent signal reports from Arliss.

This is not the only case of long path reported during Cycle 23. In April 2002, stations in south Florida, including Tom, K4SUS, and Bruce, K2RTH, worked into Taiwan via long path by keeping their antennas pointed southeast (about 150 degrees from their location). This is truly mind-boggling as the stations on both sides of the QSOs were located north of the geomagnetic equator!

Going the Distance

There remains some mystery to long-path F_2 propagation on 6 meters, but there is no doubt that there's actually a combination of at least two modes involved. Most likely TEP and some sort of F_2 layer paths combine to cover these distances. In fact, because of the distances achieved (sometimes greater than 10,000 miles), the path may involve one or two chordal hop-type paths where the signal is extended in the *F* layer rather than an intermediate ground reflection between certain points of entry. Long path usually involves stations located within the vicinity of the equatorial zone or near the tilt angle or entry point of the zone. Observations have shown relative consistency of the signals, with generally no rapid fading, so it appears that there are no *E*s links involved in long-path 6-meter openings.

Indeed, long-time observers of this mode, such as Jim Kennedy, K6MIO/KH6,[14] believe that long-path skip on 6 meters may involve both F_2 and TEP links and that the signal

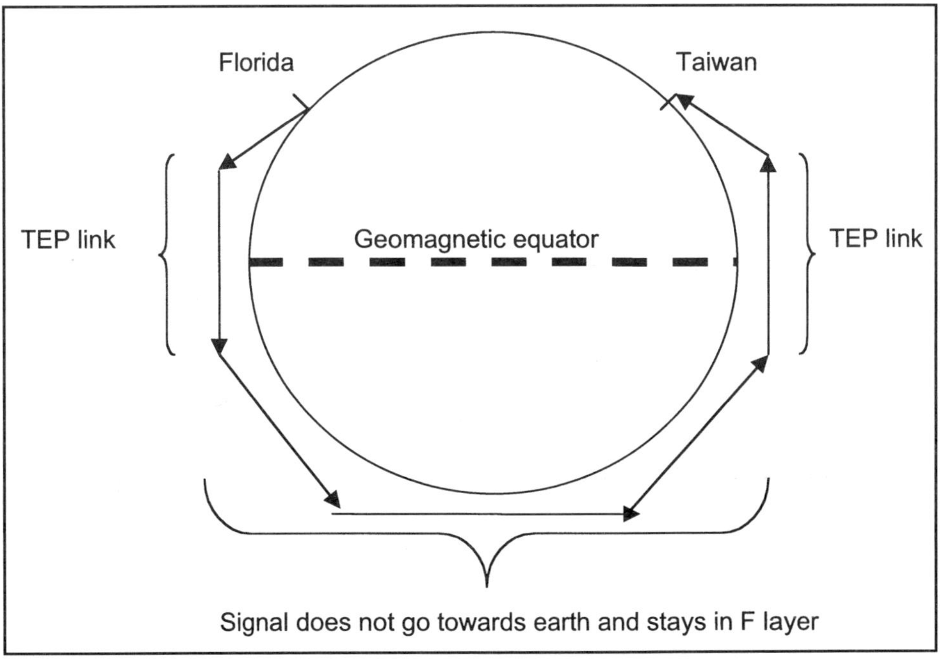

Figure 8-2. Long-path 6-meter propagation can be modeled as shown in this figure. Note that the signal stays in the F *layer of the ionosphere and at no point between the two stations does it come down to earth.*

may never reach earth in between the two locations involved. This is substantiated by the fact that there appears to be low loss on the signals and that QRP contacts have been made. This would indicate that the signal is coming from one link and skimming off the top of the F_2 layer (in chordal-hop fashion) at the next link (see **Figure 8-2**). Each link in the path (even the tops of the formations) must have high Maximum Usable Frequencies (MUFs) capable of reflecting 50-MHz signals.

It may well be that any chordal-hop-type path going over the geomagnetic poles would behave quite like TEP, where heavy ionization in the F_2 layer is involved. This may explain why this path appears more during the equinox period, with the month of April historically a good bet. As F_2 plays a part, long-path propagation will only be seen during the peak years of the sunspot cycle.

The TEP+*E*s combination appears to occur more in the Americas than anywhere else in the world. This is because the geomagnetic equator dips heavily here (from 5 to 10 degrees lower than the geographic equator). Thus, when TEP comes into the 20-degree North latitude mark, there is still a significant amount of area that is above in the northern temperate zone for an *E*s link to occur to provide the hookup. Elsewhere in the world, the geomagnetic equator is much closer to the geographic equator and TEP arrives in the higher latitudes of the north temperate zone. This leaves only a certain number of places, such as Scotland and the Scandinavian countries, that can be linked via *E*s to the TEP opening.

Signals from Australia and New Zealand can occasionally make it into Florida on 6 meters during the high sunspot count years, usually during October or April and usually later in the day. This particular signal path behaves like TEP even though the signals are not exactly perpendicular to the equator, as in traditional TEP. Most likely the mode is also similar in that the source of its ionization is near the geomagnetic equator. This significant distance could even signify a dual F_2 skip from the Australia area that appears to behave like TEP.

But what about a long-distance path where the northeast United States can work into Australia? This is a distance of over 12,000 miles, but it has been accomplished!

A case in point is when Frank, AA2DR, on Long Island worked Ron, VK4BRG, in Queensland, Australia, on October 29, 1991. Ron was working into the southern United States at around 2100Z when an *E*s link brought his signal into the northeast. Thus this particular path probably involved three different links: *E*s+TEP+*E*s.

It should be noted that the TEP link is not the standard north-south path TEP but a skewed F_2 path that crosses the equator.

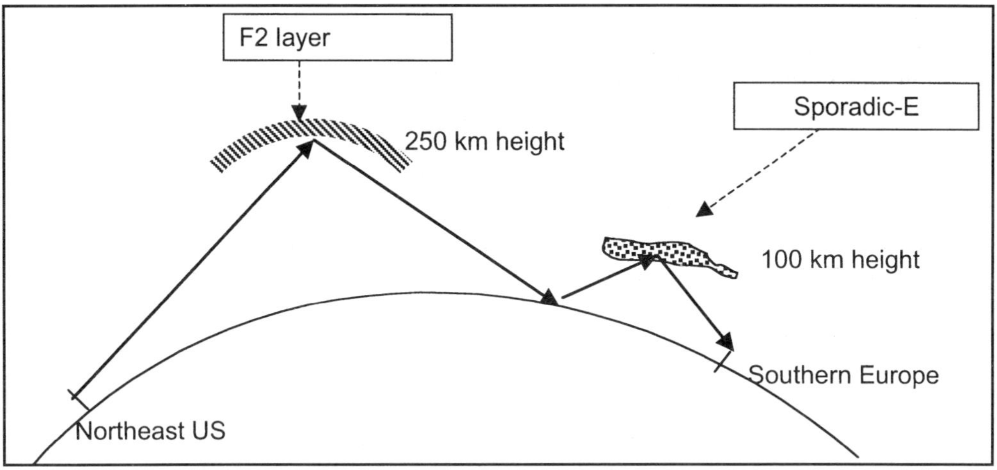

Figure 8-3. This particular combination allowed for stations in Southern Europe in the Mediterranean area to be worked by stations in the Northeast United States. WB2AMU was able to work several stations in the Mediterranean area of Europe, covering Spain and France along with rare 6-meter DX, such as ZB2EO and ISØGQX. This opening occurred well after the local sunset of the European stations, from 1700Z on, so it appears that the most likely combination was that there was an Es *skip linking up with the* F_2 *path.*

This skewed path may actually be two F_2 skip hops, judging by the higher activity surrounding the equatorial zone. So there is also the possibility that the path is actually $Es+F_2+F_2+Es$, though there is no way to know for sure. The only certainty is that there were at least two different propagation modes involved (*E*s and some form of *F*-layer propagation, possibly double-hop F_2 or TEP. The individual distances covered by either of these combination links added up to a total of 12,000 miles. And, by the way, Frank repeated this feat late one afternoon in early November, 2002, when he contacted VK2FHN.

*E*s can appear occasionally during the winter months. So during the high sunspot count years, even winter *E*s and F_2 combinations are possible. On January 4, 2002, in the early afternoon, Ken, WB2AMU, worked several Mediterranean stations, including Spain and France, along with rare 6-meter DX, such as ZB2EO in Gibraltar and ISØGQX in Sardinia. This opening occurred about 1700Z, well after the local sunset of the European stations, so it most likely was an *E*s skip linking up with the F_2 path (see **Figure 8-3**).

At certain times of the year, during the sunspot peaks, double-hop *E*s can couple with F_2 propagation on 6 meters. This can allow stations in the northeast United States to work into Central Africa. One such instance took place in May 1999, when Mick, W1JJM, worked a 5H3 station. Mick was hearing Spain and the

Azores via double-hop *E*s at the same time that some of the European stations were working into Africa via F_2 skip. F_2 is not normally present in May but may sometimes be induced by high geomagnetic activity, as was the case here (see **Figure 8-4**).

Other example of double-hop *E*s coupled with F_2 occurred when Bob, VE1YX, in Nova Scotia worked FH/TU5AX in Mayotte off the east coast of Africa on June 6, 2000, and Dave, K2SIX, worked 9J3BO of Zambia in the early afternoon of July 11, 1999. These instances show that it pays to be vigilant on 6 meters during the heart of the summertime *E*s season in high sunspot count years!

Another event that happens quite frequently during the summertime *E*s season is a combination meteor scatter (Ms) and *E*s contact. This is a difficult scenario to define conclusively but it basically takes the form of a very strong signal that's heard for 10 to 30 seconds on 6 meters and then nothing is heard again. A long meteor burn may last for five to 10 seconds, but a 30-second duration suggests an *E*s opening. What appears to be happening is that there is an *E*s cloud present, but its density will only reflect an MUF less than 50 MHz. The presence of meteors passing through the *E*s cloud may actually induce increased ionization of the formation, and thus support a higher MUF.

Yet another combination that can occur is F_2 and Au. A case in point occurred on January 7, 2002, when Tim, N1RZ (now W4TRH), and others in northern New England were able to work into the Scandinavian countries shortly after noon from his QTH in New Hampshire. Starting at 1652Z, Tim worked OH1XT in Finland, several others in Sweden, Finland, Norway, and OY9JD in Faroe Island in the north Atlantic. All the signals

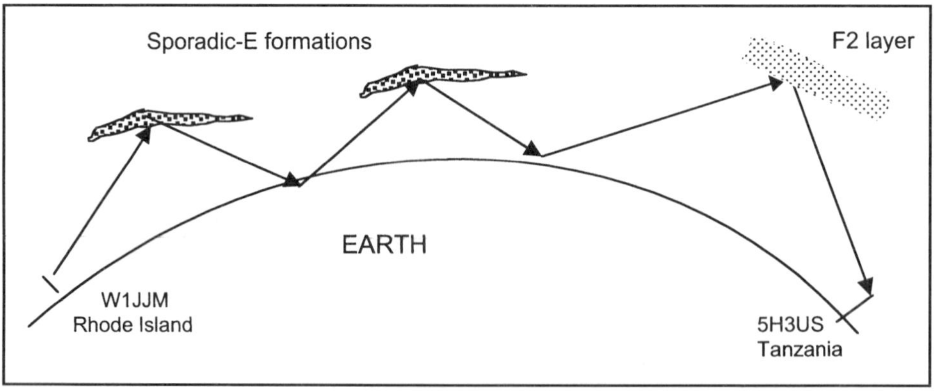

Figure 8-4. The above occurrence took place at 1840Z on May 26, 1999. Two-hop Es *was observed into the Azores/Spain area of Europe at the same time that Europe was working into Africa via* F_2 *skip.*

Table 8-1. Six-meter Propagation Mode Occurrences During Peak Sunspot Years

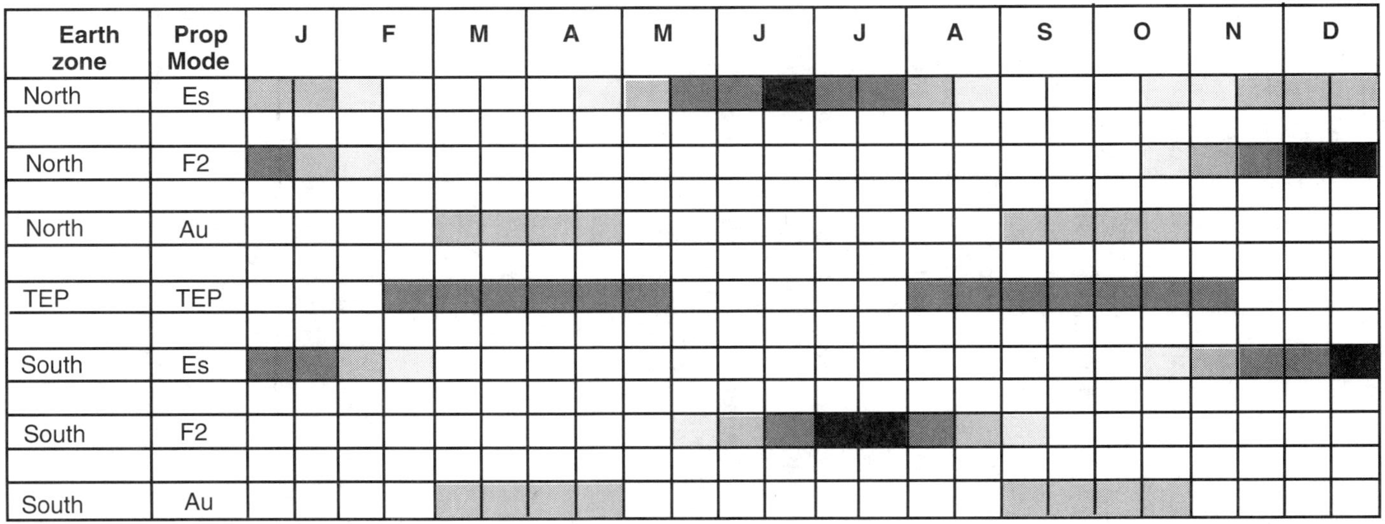

KEY (percentage of days with propagation activity)

75–100% 50–75% 25–50% 5–25% less than 5%

had the distinct aurora tones, so the combination was likely an Au path in northern Europe that linked up with the F_2 path into the United States. The opening lasted until 2 p.m. local time for Tim, when he worked OZ4VV in Denmark with Au signals. This combination happens on occasion during the peak of the F_2 season of the higher solar flux years, but primarily just between stations in the higher latitudes on both sides of the opening (above 40 to 50 degrees north).

There have been examples of the Au+F_2 combination working in the reverse, where the aurora was on the North American side. This was the case in the contact that W1JJM (now W1JJ) in Rhode Island made with GIØOTC on December 27, 2000. Here W1JJM heard some stations in Canada working GIØOTC via F_2, while *he* heard those same stations via Au backscatter. He was able to work GIØOTC through the combination of the signal coming into Canada via F_2 and then to Rhode Island by Au backscatter. It's truly amazing to see the kinds of paths that a radio signal can travel on our ham bands, and 6 meters appears to be a great band for observing these unusual combinations!

These long-range openings on 6 meters are not usually very long in duration, perhaps 20 minutes on average. While the TEP mode may last for a while between locations like Florida and Argentina, *E*s is always unpredictable as to duration. So how long a TEP+*E*s combination (with the latter linking other areas in the United States to Florida) might last is certainly anybody's guess.

It is noted that many of these combinations of propagation modes occur during the higher sunspot count years of the 11-year solar cycle, because modes such as F_2, TEP, and Au show up more often then. Such combinations of these modes can possibly occur during the low count years under rare circumstances. One such circumstance would involve equatorial ionization activity that occurs after an Au opening where TEP follows.

Table 8-1 summarizes the modes of propagation that can appear on 6 meters throughout the year during the peak sunspot years. By going up and down each column, one can see the combinations that are possible under the right circumstances. As previously mentioned, the months of October, November, April, and May provide the best chances for combinations of modes to occur. These combination-type openings don't usually last very long, so vigilance during these months is required! Persistent monitoring of the band or an Internet site dedicated to 6 meters will help you keep on top of such events.

The Next Challenge

Can combination mode events occur on 2 meters? In theory, yes, but certain conditions must be present. F_2 is not possible on

2 meters because the MUF does not reach that high. TEP has appeared on 2 meters for certain stations in the Caribbean, and certainly an *E*s link, or even a meteor link, could make a contact possible for stations in higher latitudes. Because of the very short duration of such events, digital modes or high-speed CW appear to be the only way this could be accomplished.

In fact, the Irish Radio Transmitters Society has issued a challenge in recent years. It will offer an award, called the Brendan Trophy, for any amateurs who can complete a QSO on 2 meters via terrestrial means only (not using space/satellite modes) between mainland Europe and North America. There have been a number of attempts, with the approach being to make the attempt in the summer when double-hop Ms, or an *E*s skip plus Ms, is possible. One simply can't predict when such a combination would occur, so efforts of week-long duration have been the norm.

For 220 MHz and 432 MHz, Ms contacts are even more difficult, and Au rarely reaches this high a frequency. Six meters remains the VHF band with the best chances for combination events, with 2 meters seeing rare cases of combination events.

In Summary

It was pointed out in this chapter that the best time for these different combinations of propagation modes to occur is during the higher sunspot count years of the 11-year solar cycle. That is because modes like as F_2, TEP, and Au show up more frequently and can combine on occasion with the *E*s mode (which shows up throughout the entire sunspot cycle). Such combinations can possibly occur during the low count years, but only under certain circumstances, such as when high geomagnetic activity occurs and both Au and TEP events result. It is by these interesting combinations that unusual DX can occur on the 6-meter band.

CHAPTER 9

OVERVIEW OF WORLDWIDE 6-METER PROPAGATION

At this point in the book, the reader should have a good knowledge of how the different 6-meter propagation modes work and when they generally appear. Understanding and observing these modes on a band like 6 meters should serve as a good working background on predicting how they will behave. It should be emphasized that some of these modes, such as sporadic-*E* (*E*s) and aurora (Au), rarely appear in force on the HF frequencies, particularly 20 meters and below. It helps for the reader to actually observe these modes on 6 meters when making contacts.

In this chapter, we'll discuss how this knowledge can be helpful to operators, both in their home locations and when they travel to other areas of the world as part of DXpeditions.

As was described in the previous chapters, the earth has three different zones: aurora, temperate, and equatorial. This unique geophysical setup governs the behavior of certain propagation modes, such as *E*s in the temperate zones and Au in the aurora zone.

Because of the differences among the zones of earth with regard to the geomagnetic poles and geomagnetic equator, what you hear on 6 meters during the course of a year and sunspot cycle will depend on where you are. It is also important to realize that these various propagation modes differ in frequency of occurrence depending on your location. Six-meter propagation conditions, particularly, will vary through different areas of the world. You should be aware of this, especially for DXpeditions, if you hope to make long-range contacts on six.

Location, Location, Location

Table 9-1 provides a general summary of propagation conditions in some of the areas of the world based on both amateur radio observations and scientific studies.

Looking at the table, it's easy to see why certain parts of the world do so well with some of the operating awards. For example, because of the decent *E*s conditions during the summer in Europe, it's reasonably possible for many stations to work over 50 different countries in a single season! In fact, many European

Table 9-1. Summary of 6-Meter Propagation Conditions Worldwide

Location	***E*s (summer)**	**Au**	**TEP (equinox)**	**Tropo**	**F_2**
USA – South (below 25° N)	* * *	0	* * *	* * *	* * *
USA – Northeast	* * *	* *	(see notes)	*	* * *
USA – Midwest	* * *	* *	(see notes)	*	* *
USA – Southwest	* * *	0	*	* *	* *
USA – Northwest	* * *	* *	(see notes)	* *	* *
USA – West coast	* * *	0	* *	* * *	* *
USA – Alaska	* * *	* * * *	0	*	* *
USA – Hawaii	* * *	0	* * *	* * *	* * *
Canada – Eastern	* * *	* * *	0	*	* *
Canada – Central	* * *	* * *	0	0	*
Canada – Western	* * *	* * *	0	*	* *
Japan	* * * *	0	* * *	* *	* * *
Europe – Western (UK, France)	* * *	* *	(see notes)	*	* *
Europe – Northern (Scandinavia)	* * *	* * * *	0	*	* *
Europe – Southern and Mediterranean area	* * * *	*	* * *	* *	* * *
Caribbean	* * *	0	* * *	* *	* * *
South America (above equator)	* * *	0	* * *	* *	* * *
South America (below equator)	* * *	0	* * *	* *	* * *
West Africa	* * * *	0	* * *	* *	* * *
South Africa	*	0	* * *	*	* *
Australia and New Zealand	* * *	*	* * *	* * *	* *

KEY TO TABLE

* * * *	Excellent activity
* * *	Moderate activity
* *	Occasional activity
*	Rarely occurs
0	Never occurs

Notes:
1) F_2 occurs during peak of sunspot cycle.
2) TEP into Northeast, Midwest, and Northwest U.S. and Western Europe requires an *E*s link.

stations have achieved the 6-meter DX Century Club (DXCC) in recent years. Another "hot spot" is Japan. Here you find good *E*s, transequatorial (TEP), and F_2 conditions that make it possible for an active 6-meter operator to achieve 6-meter DXCC—and hundreds have to date! However, it's harder for stations in Europe and in Japan to achieve Worked All States on 50 MHz, whereas this goal is within reach for a number of U.S. stations.

South Africa has the poorest incidence of *E*s of any area in the world. This is believed to be related to geomagnetic factors (remember, the earth is not a uniform sphere). But there are decent TEP conditions between South Africa and southern Europe. Also, there are great TEP conditions between Australia and Japan, as Jack, N6XQ, found during a visit to Australia. And the list goes on.

There is no one perfect area of the world where more than four major propagation modes regularly appear in force on 6 meters. For example, a station located in southern Florida would experience TEP on that band during the equinox period, but would rarely have the privilege of hearing aurora signals, except in major openings. Likewise, a station in the northern United States would have plenty of aurora openings, but would only be able to experience TEP if an *E*s link was present (as we learned in **Chapter 8**). No one location has a monopoly on all the VHF propagation modes.

Another important thing to note is the apparent fickleness (or narrowness) of propagation, even when it is open for the same general geographic area. For example, one may not be able to hear a station coming into New York from Florida via *E*s, while another ham 30 miles away can hear and work the station. This change in propagation coverage can sometimes be noticed while going mobile. This shows that certain modes of propagation, such as *E*s and Au, may not be uniform inside the formation and that radio wave signals may be reflected differently. Thus, if a station in New York sees reports of Europe being worked by stations in the New England states, that station may not be able to hear anything, even if he has a decent setup. In the end it's combination of propagation and, of course, luck. It just may not be going your way when this happens, but every dog has its day, so don't give up and try again the next day.

Yet another important characteristic of *E*s propagation (as discussed in **Chapter 3**) is that the higher in latitude one travels during the time of the summer solstice, the better the chances for encountering *E*s. For example, the northern latitudes around the time of June 21 see many more hours of sunlight and more opportunities for *E*s during the day, even shortly after dawn. Stations that have traveled to northern Canada, say to Zone 2 in Northern Quebec or even higher, during the summer have noticed this.

In recent years, a number of DXpeditions to rare countries have been including 6 meters as one of the bands to be worked, in addition to the usual HF band operation. Sometimes such expeditions will have a dedicated 6-meter operator to monitor the band full time; however, some expeditions have fewer members, so 6 meters is a part-time band. The more successful operations will have a full-time operator on the band because of the fleeting propagation that may occur. For example, the Galapagos Islands (Prefix HC8) in the Pacific Ocean has been the subject of many DX operations that take place during the CQ World Wide DX Contest. During the peak sunspot count years, Jon, NØJK, joined a few of these efforts as the full-time 6-meter operator, while other operators concentrated on HF. There is also a full-time beacon on one of the islands, which has been heard frequently in much of the United States on 6 meters during F_2 openings. Jon was able to take advantage of this same F_2 path during his November 2001 operation and worked quite a few U.S. stations on six. He has conducted many such trips to different islands in both the Atlantic and Pacific Oceans, with 6 meters as the band of focus. As part of the preparation for these trips, Jon (and others like him) will take careful note of the potential propagation modes that may be available at the time of travel.

Another dedicated 6-meter operator who conducts similar trips is Arliss, W7XU. During his trip to Easter Island in April of 2001, Arliss made many contacts into the United States via F_2, TEP, and TEP+*E*s links. But what was even more interesting is that Arliss discovered a long-path *F*-layer link that connected his location with India, to the point that QRP contacts were being made! It shows the true value of DX trips like these, which can lead to the discovery of unique propagation modes or previously unknown paths!

Dedicated 6-meter operators will often conduct 6-meter-only expeditions to work DX countries and rare grid squares. Suffice it to say that it's mostly during the summertime *E*s and tropo season that visits are made to islands off the shores of the United States, such as Bermuda and the Caribbean islands, that are still somewhat rare on 6 meters.

For example, Bermuda, while somewhat common as a country on the HF bands, became rare for several years on 6 meters towards the end of the last century. Limited activity took place there on 6 meters during the 1990s by visiting hams, including W3EP and WB2AMU. By 2003, there was more concentrated 6-meter activity, in part thanks to visiting hams like Jon, NØJK, Chris, W3CMP, and George, GM4COK. During recent trips they had left behind antennas and radios for local stations to use, so the local amateur radio population has been more active on six.

As part of his visit to Bermuda, W3CMP also provided the support for a full-time beacon (VP9DUB/b) to be put into action on 50.025 MHz. This, in turn, provided valuable information on many *E*s openings between Bermuda and parts of the United States that would have otherwise gone unnoticed. Indeed, it has also been noticed that there seems to be decent propagation to Europe during the summer months from this location as well. In addition to the 6-meter activity, some 2-meter work was attempted from Bermuda with the hope of finding any existing tropo paths to the United States.

Two for the Road

Even though we are discussing 6 meters in this chapter, we want to mention that there are also 2-meter-only expeditions to other locations. In **Chapter 2** we discussed the weather conditions that lead to excellent tropo propagation for certain parts of the world. And the beauty of enhanced tropo conditions is that one can take advantage of them at certain times with just a small HT radio on the 2-meter and 432-MHz FM band.

Of course, notable 2-meter DX trips of recent years have been undertaken in the attempt to transmit the first 2-meter signal across the Atlantic Ocean from North America to Europe using terrestrial propagation. Some attempts have included high-power setups operating in the coastal areas of the United States and Newfoundland on one side and setups in the Irish coastal area on the other side. There is a tremendous amount of luck and good timing involved as some sort of two-hop skip involving *E*s, meteor scatter (Ms), and tropo would be needed to make this particular path happen.

In Summary

The earth is not perfectly uniform in terms of field strength, shape, and weather conditions. As a result, propagation conditions can vary significantly on the VHF bands (particularly 6 meters), depending upon where on the planet your station is located. By being aware of the types of propagation modes available at particular times of the year, one can take advantage of the optimal times for operating VHF when visiting other areas of the world.

CHAPTER 10

SPACE BOUND SIGNALS

In this chapter, we'll provide a general overview of a couple of the space-oriented modes on the VHF bands and how the ionosphere affects these signals. Specifically, we'll look at earth-moon-earth (EME) and satellite modes. Essentially, in these modes, radio waves leave the earth, passing through the ionosphere rather than being reflected back.

In previous chapters, we discussed most of the ionospheric modes of propagation, including meteor scatter (Ms), which is a sort of space mode as the meteor particles travel a long distance in space before finally burning up in the earth's ionosphere. However, hams experience Ms primarily as a propagation mode that takes place in the *E* region of the ionosphere and not as a true space mode.

Working EME

The moon is approximately a quarter of a million miles away and is pocked with craters, which makes it a non-perfect reflector. You would be lucky to obtain five percent of your signal back from the moon because much of the signal's energy is scattered through the heavens; in fact, path loss on the 2-meter band is as great as 232 dB! There is about a 2-dB change between perigee (the closest point to the earth) and apogee (the furthest point from earth). Yet the moon offers amateur operators another target to practice reflecting signals off of. Typically, you would need multiple Yagi arrays, rotators, and pre-amps to get started. Then, because the signals are to be reflected off the moon, you would have to learn how to track the moon in azimuth and elevation from moonrise to moonset.

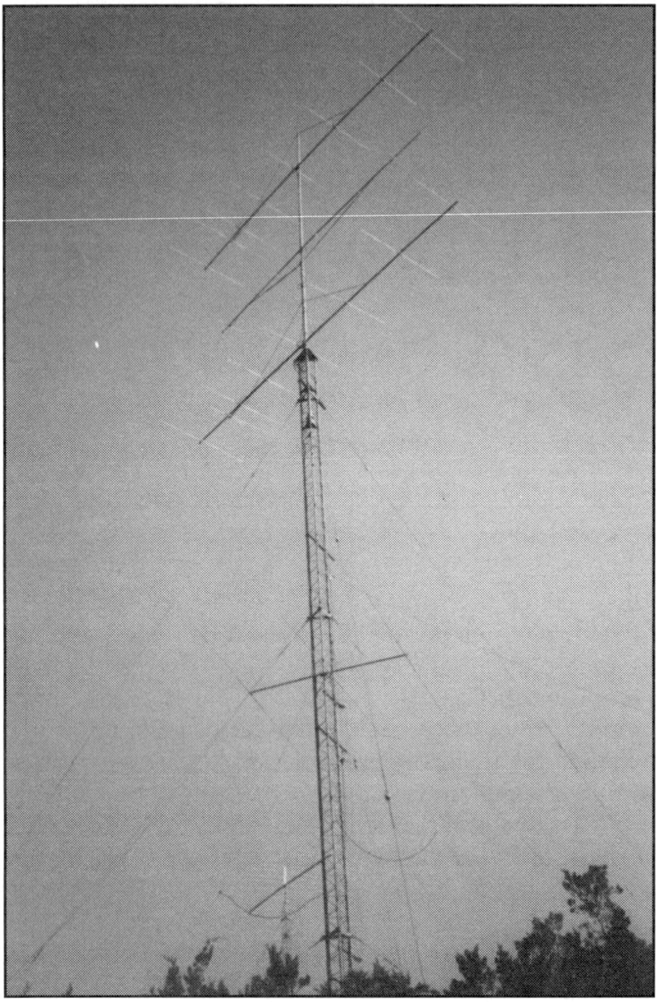

Photo 10-A. This stacked pair of eight-element Yagis on a tower is actually one of the simpler setups that may be used for EME work. Other EME operators may have as many as four or eight multi-element Yagi antennas for their arrays and will also have sophisticated rotators for better tracking with the moon. (Photo by Gordon West)

Finally, you would need a great deal of operating skill because of the associated weak signals.

EME work is done on 50 MHz, 144 MHz, 432 MHz, and 1296 MHz, with 144 MHz being probably the most popular frequency for this type of work. Two meters seems to be the ideal frequency in terms of size of antenna and available equipment. As there is an estimated path loss of about 232 dB, it's generally accepted that to be successful on this band a station of four 2.2-wavelength Yagis, representing 20 dBi, would be needed, along with

1000 watts of power. However, some stations with a single Yagi and 100 watts have been successful in working the "big guns" off the moon at local moonrise/moonset, taking advantage of a few dB of ground gain. There have even been a few single-Yagi to single-Yagi EME QSOs by stations using JT-65, one of K1JT's digital modes designed specifically for EME.

The path loss on 432 MHz is about 9 dB greater than it is on 2 meters because of the "antenna" effect. In essence this means that a dipole required for this band would be one third the size of one used for 2 meters, so it will only capture one-ninth the power of the 2-meter dipole. Therefore, to make up for this difference, a larger array is needed to get the equivalent "capture area." Thus one could end up with four 6-wavelength Yagis to meet the minimum gain requirements of 24 dBi.

There are more hams doing 2-meter EME than 6-meter EME as the latter requires a larger antenna to be effective, typically with boom lengths of up to 70 feet. Since long-range DX can be worked on 6 meters with moderate success at certain times of the year and solar cycle, much of the emphasis of EME work is placed on 2 meters and higher.

EME contacts are typically made using CW, but sometimes conditions allow SSB to be used. Similar to Ms, there is an operating procedure typically followed when making contacts. On 2 meters, the eastern-most station transmits for two minutes and then listens for two minutes for the other station. A successful exchange requires the callsign and a signal report. Typically the number zero is used for a report, unless conditions are really strong enough to use a conventional RST report format. On 432 MHz, the western-most station starts first, and the time period is typically two and a half minutes. A schedule may be set up for an hour block of time until it is called off.

In addition to using multiple antennas, you may need to approach the legal limit of 1500 watts if you're trying to hear your own echoes off the moon with a single or pair of line-boom Yagis (see **Photo 10-A**). With more exotic, higher-gain antenna systems, your power requirements would be less. You'll need a mast-mounted pre-amp and narrow filters in the receiver to help minimize the amount of signal noise, which may be enormous on a hot receiver picking up cosmic noise sources, including sun noise saturating the receiver. Also, it's best to use hardline-style feedline and to keep lengths to a minimum (75 feet or less is recommended) to keep down line loss.

Yet another few dB of signal loss results from something called libration fading. This refers to a slight rocking of the moon on its axis, which causes slight inconsistencies in reflections. It is both unpredictable (without an extensive astronomy background) and hard to compensate for without huge antenna systems.

VHF signals for EME and satellite modes are somewhat affected by the ionosphere as they pass through it. As a radio wave passes through the ionosphere, it will be rotated in polarity by an amount depending upon the ionization level that it passes through. This is known as Faraday Rotation and is depicted in **Figure 10-1**. The signal will be further rotated on the path back. The rotation of the signal will always be changed at a speed that depends upon the frequency used. For instance, the speed is faster at 50 MHz and slows down as you climb in frequency.

The loss may be significant—in the area of 25 dB—due to Faraday rotation during the trip to the moon and back down to earth. When it passes through the ionosphere, it may rotate up to 90 degrees out of phase to your antenna system. This rotation may slowly change over a few minutes, so two hams participating in a moon bounce effort may simply have to "wait it out" until like polarizations are achieved.

As EME signals tend be very weak, even with the use of high-gain antennas and preamps on the receiver, the key to successful EME operation is taking the necessary steps to reduce noise, both external and system-generated. For instance, it may be better to attempt EME at night because of reduced manmade noise; on the other hand, nighttime dew on power lines can mean an increase in noise!

As you can see, EME, requires a lot of dedication and effort. There have been a lot of pioneering efforts by these hams in discovering the different aspects of EME.

Working Satellites

Amateur radio has been a major force in the satellite field ever since December 12, 1961, when the first amateur satellite, OSCAR I, was launched piggyback with Discoverer-36, a United States Air Force satellite. In *The Satellite Experimenter's Handbook*, Martin Davidoff wrote:

> OSCAR I was an overwhelming success. More than 570 amateurs in 28 countries forwarded observations to the Project OSCAR data reduction center. The observations provided important information on radio propagation through the ionosphere, the spacecraft's orbit and thermal satellite design. The OSCAR I mission clearly demonstrated that amateurs are capable of (1) designing and constructing reliable spacecraft, (2) tracking satellites, and (3) collecting and processing related scientific and engineering information. Because of its low altitude, OSCAR I only remained in orbit for 22 days before burning up as it re-entered the earth's atmosphere.

The success of OSCAR I led to the formation in 1969 of the Radio Amateur Satellite Corporation (AMSAT), a worldwide group of amateurs who share an interest in building and launch-

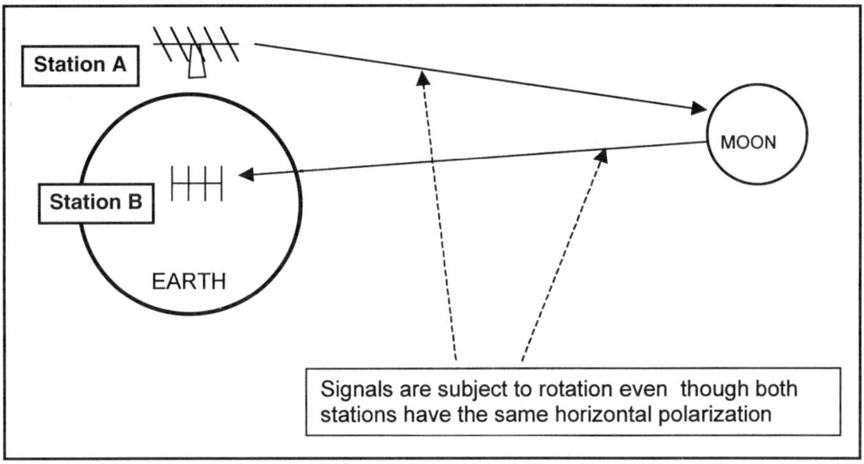

Figure 10-1. A depiction of the effects of Faraday rotation on signals traveling from the earth to the moon and back to earth.

ing amateur satellites and then communicating with each other through them. Thanks to AMSAT and its subsequent launches, we constantly have a view of ham-built satellites that may be worked with modest antenna systems, including two satellites that may be worked using the FM mode with any full-duplex, dual-band handheld transceiver and the recommended telescopic whip. Of course, any portable dual-band Yagi will work great in tracking these FM satellites as they come and go several times a day over your area.

The more distant satellites that may work on microwave frequencies will certainly require a more elaborate home antenna system, along with home-brew down-converters and up-converters, but the signal levels are well above the noise and easily permit SSB and CW contacts. To help you out there are numerous satellite tracking programs, personal data devices, and simple programs for the home computer that constantly show you where the ham satellites are as well as their expected range to other stations with a mutual view.

It should be pointed out that this satellite tracking methodology has been useful in the area of contacts made between astronauts in the space shuttle or space station and amateurs on earth. VHF bands, such as 2 meters, are used to bring the signals from the astronauts in space into classrooms around the world as they pass overhead at the right time.

These satellites or spacecraft are above the ionosphere, so we're not dealing with issues of ionospheric propagation, but rather with the impact of the ionosphere on radio waves, such as signal loss (just like in EME work). There are some effects of signal absorption and possible rotation (the distances are significantly less going to a satellite as compared to the moon). Thus, it becomes important to accurately determine the time of overhead flight and path a satellite or spacecraft will take for optimal signal strength.

Follow the Bouncing Wave

Radio waves at VHF and UHF can also be bounced or scattered off a variety of unusual things, and while we won't go into any detail on them because they're pretty uncommon, we thought we'd close with just a brief mention of some of them.

Perhaps the easiest to visualize is building scatter. If two stations that cannot contact each other via a direct path can both see a large building, then both stations can aim their antennas at the building and can often complete a communication path by bouncing their signals off of it. This tends to work best at 70 centimeters and up, into the microwave frequencies, and it's why 70 centimeters often gets through better than 2 meters in downtown urban areas, as the signals bounce off of buildings. The same effect can be achieved with airplane scatter, which is only effective near airports with lots of traffic, since airplanes don't stay in one place for very long.

Another phenomenon you may encounter is multipath, in which reflections off multiple large objects (such as buildings in a downtown area) can affect VHF/UHF propagation ranges. It's usually evidenced as rapid fading, but what's actually happening is that the receiver is picking up the signal and several reflections of the signal simultaneously. Since the reflections have traveled a different distance than the main signal, they arrive out of phase. Those 90 or 270 degrees out of phase cancel each other out, while those 180 or 360 degrees out of phase amplify the signal, so you generally have a quick succession of peaks and valleys.

Up in the microwaves, there are also such things such as rain scatter, snow scatter, and knife-edge diffraction, all of which are beyond the scope of this book, but are fascinating to learn about if you choose to dig deeper into the world of VHF/UHF propagation.

In Summary

Chances are you wouldn't automatically think of our ionosphere as a major influence on moon or satellite communications, but studies have proved that the ionosphere may become highly disruptive to weak-signal propagation. In addition to the scientific studies conducted on how the ionosphere impacts position-finding with global positioning system receivers and commercial satellites, hams have discovered much about the effects of the ionosphere on EME and satellite operation. The information gained by hams using these modes continues to be a very important contribution!

CHAPTER 11

EPILOGUE: AMATEUR RADIO AND VHF PROPAGATION

We have provided a lot of practical information about VHF propagation in this book. Because of increasing VHF activity, the knowledge base is continually expanding as more contacts are made using the different modes of propagation. It is hard to believe, but each contact made by a ham using a particular VHF mode can represent a major step in understanding certain aspects of that mode, whether it's aurora (Au) or F_2 propagation. **Table 11-1** summarizes the propagation modes that are typically seen on each of the VHF bands.

Amateur Radio's Contribution to Understanding VHF Propagation

Amateur radio operators have contributed greatly to the radio exploration of space and the upper atmosphere. In these pages, we have highlighted amateur radio as a focal point of discovery for the sporadic-*E* (*E*s), Au, and meteor-scatter (Ms) propagation modes. We have also highlighted amateur radio's participation in scientific studies of the major propagation modes that have led to increased understanding of their behavior and characteristics.

When we examine many of the major space communications breakthroughs, it's very likely that we'll find several ham operators on the commercial exploration team who contributed greatly. In fact, there are many areas outside the hobby where hams have contributed.

For example, the work done by hams using satellites and EME has added to the knowledge base that led to the development of various types of satellites as well as the global positioning system (GPS). GPS relies on radio signal propagation down to

Table 11-1. Major Propagation Modes by Frequency for Northern Temperate Zone

Frequency	Tropo	*Es*	Au	Ms	F_2	TEP
50 MHz	X	X	X*	X	X*	X*
144 MHz	X	+	+	X	-	-
222 MHz	X	-	-	-	-	-
432 MHz	X	-	-	-	-	-
902 MHz	X	-	-	-	-	-

KEY
Tropo = Tropospheric propagation
Es = Sporadic-*E* propagation
Au = Aurora propagation
Ms = Meteor-scatter propagation
F_2 = F_2 layer propagation
TEP = Transequatorial propagation

X = observed several times yearly (except those entries marked with * which require high solar activity)
+ = observed one or two times a year
- = very rare occurrence

extreme levels of accuracies, and the variations within the ionosphere can make a big difference in the effectiveness of the mid-earth-orbit GPS satellites. In the area of geosynchronous satellite "repeaters" for television and streaming satellite radio, atmospheric attenuation and ionospheric absorption are a major concern. An in-depth understanding of propagation is also needed to establish receiver antenna requirements for land stations, as well as for viewers and listeners out on boats and in airplanes.

From our experiences as VHF operators, we are also aware of the detrimental effects that certain phenomena in the ionosphere have on satellites. We know that a major solar storm or particles from a coronal mass ejection can damage the electrical circuitry of satellite instrumentation. We also know that, on occasion, these satellites may be prone to damage from meteor showers. The observations that VHF operators make concerning these phenomena are helpful to the users of commercial and scientific satellites.

As we detailed in **Chapter 5**, hams were pioneers in the use of high-speed CW in Ms work, and contributed to the overall understanding of the Ms mode. This work has led to commercial high-speed Ms stations that continue to blast away in the wilderness, successfully establishing data paths that would be impossible using conventional VHF land propagation predictions. It's also notable that most of these commercial operations usually have active ham operators on their staff! Note, too, that many astronauts have obtained their amateur radio licenses so that they can operate on the VHF and UHF bands from space.

VHF Propagation, A Summary

As we have seen in this book, the sun is the key to many of the VHF propagation modes. During the day, the sun bombards our ionosphere with ultra-violet radiation that causes the *D*, *E*, and F_1/F_2 layers to form and make their reflective and refractive "magic." At nighttime, these ionospheric layers begin to merge to form the single *F* layer, as chemicals recombine after a day in sunlight.

During the peak and recession of the solar cycles, the sun also gives us some startling "space weather." The charged particles from coronal mass ejections impact the top of the ionosphere (known as the plasmasphere) and, eventually, our ionosphere. As we learned in **Chapter 4**, an aurora will result and VHF radio conditions will get very exciting. The 6-meter band may pop open for hours, and 2 meters may follow as well.

We also know that the sun charges up those ions in the *F* layer that allow truly long-range contacts to be made by F_2, transequatorial propagation (TEP) and long-path modes. And we know that the sun has some impact on *Es* propagation as well. Also, the sun plays a part in creating the different weather fronts that result in tropospheric propagation. So, without the sun behaving as it does, we would be left without any VHF propagation modes.

There remains much work for hams to perform and many discoveries are waiting to be made in the area of VHF propagation. Among the major observations made by hams regarding VHF propagation during the peak of solar Cycle 23 was that of long-path *F*-layer paths that allowed 12,000-mile contacts on 6 meters in the most unlikely of directions. The discovery of these long paths on six was thanks to dedicated radio amateurs operating on a regular basis from their QTHs, as mobile stations, and during DXpeditions to areas of the world where these paths are possible.

So let us keep in mind as we make use of VHF propagation modes that we are pioneers as we continue to learn new and exciting things!

APPENDIX I

RESOURCES

There are a number of resources available to hams for further investigation of some of the topics discussed in this book, including the following:

WWV BROADCASTS OUT OF BOULDER, COLORADO

Your simple HF receiver that tunes WWV can make an excellent predictor of incoming VHF and UHF band condition changes. At 18 minutes past the hour for WWV at 5, 10, 15, and 20 MHz, you can listen for a report on ionospheric conditions.

The WWV broadcast will also indicate the relative activity of solar conditions, and elevated A and K indices with "major storm" announcements would surely signal some exciting conditions on our 6-meter, 2-meter, and 432-MHz bands.

INTERNET WEBSITES OF INTEREST TO VHF OPERATORS

Solar Data And Space Weather
http://www.sec.noaa.gov/primer/primer.html
http://www.dxlc.com/solar/indices.html

Six-Meters
http://chat.dxers.info/?b=6m
http://dxworld.com/50prop.html
http://www.uksmg.org

Two-Meters
http://groups.yahoo.com/group/sidewindersontwo

Satellites
http://www.amsat.org

EME
http://dxworld.com/emelog.html
http://web.wt.net/~w5un

WSJT (Weak Signal Communication by K1JT)
http://pulsar.princeton.edu/~joe/k1jt

APPENDIX II

VHF PROPAGATION AUDIO COURSE

Thoughout these pages, the authors have done their best to explain how VHF propagation *works*, but to really take advantage of the opportunities it offers, you have to learn how the different modes of VHF propagation *sound*. That's why the authors have also put together the "Gordon West VHF Propagation Audio Course," the Audio Course that's been referenced throughout the book.

The sounds of certain modes we've examined, such as aurora, are impossible to put into words. Knowing what to listen for and how to identify their unique sounds will enable you to work these propagation modes to greatest effect. That's where this Course comes in. Narrated by Gordon West, WB6NOA, it contains many on-the-air audio examples of the different modes that were discussed.

Audio samples in this hour-long Course include the sound of meteor reflections coming from stations over 500 miles away, the first sounds of the 432-MHz beacon coming in over a record distance from Hawaii to California, and the sound of the 6-meter band before it pops open to *E* skip in the afternoon. The authors also share plenty of tips and tricks learned in their own shacks.

To order a copy of the cassette or CD, please send your mailing address, along with $10 plus $3 Shipping & Handling for each copy, in check or money order to the following address:

GORDON WEST VHF PROPAGATION AUDIO COURSE
RADIOSCHOOL
2414 College Dr.
Costa Mesa, CA 92626

Please indicate "VHF PROPAGATION AUDIO COURSE" on the check or money order, and don't forget to specify format (cassette/CD).

For quick reference, cut along dashed line and keep these tables handy in your shack or mobile setup

VHF PROPAGATION REFERENCE PULLOUT GUIDE

MAJOR PROPAGATION MODES BY FREQUENCY

(for northern temperate zone)

Frequency	Tropo	Es	Au	Ms	F2	TEP
50MHz	X	X	X*	X	X*	X*
144 MHz	X	+	+	X	-	-
222 MHz	X	-	-	-	-	-
432 MHz	X	-	-	-	-	-
902 MHz	X	-	-	-	-	-

KEY

Tropo = Tropospheric propagation
Es = Sporadic-E propagation
Au = Aurora propagation
Ms = Meteor Scatter propagation
F2 = F2 layer propagation
TEP = Transequatorial propagation

X = observed serveral times yearly (except those entries marked with * which requiare high solar activity)
+ = observed one or two times a year
- = very rare occurrence

SIX METER PROPAGATION CHART

(During years of high solar activity for northern, southern and equatorial zones)

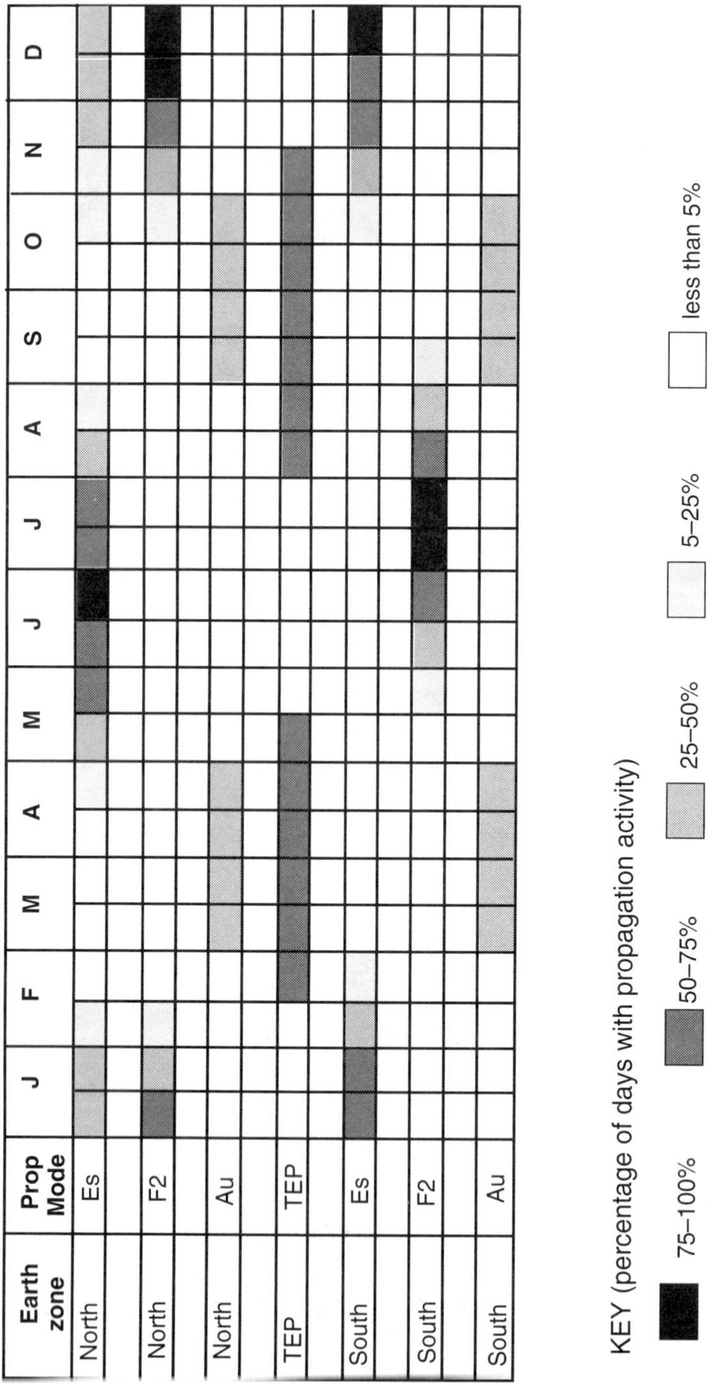

For quick reference, cut along dashed line and keep these tables handy in your shack or mobile setup

YEARLY 50 MHZ SPORADIC-E PATTERN FOR NORTHERN TEMPERATE ZONE

(Percentage of days per month that Sporadic-E is observed)

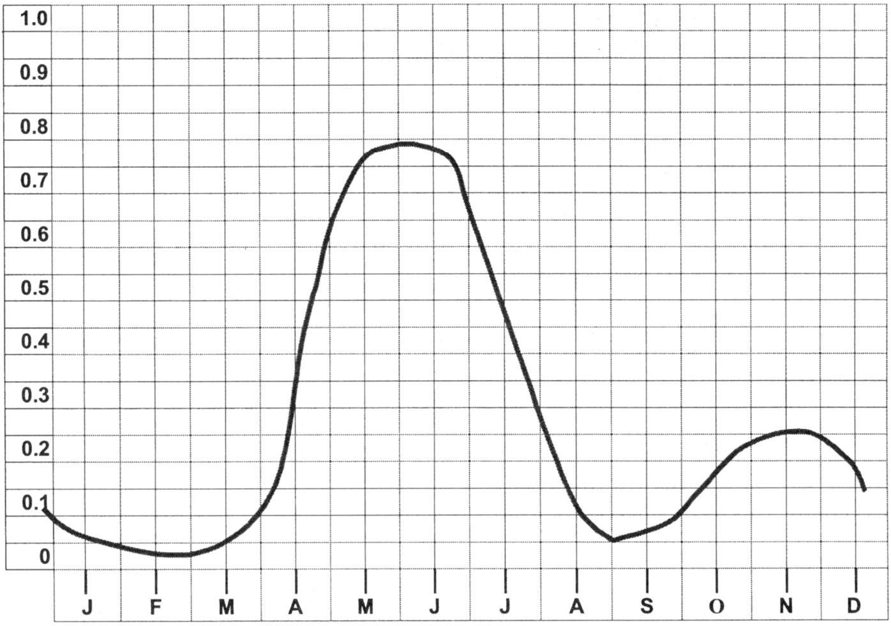

DATES OF METEOR SHOWER ACTIVITY

NAME	DATES
Quadrantids	**January 1-6**
April Lyrids	April 19-24
Eta Aquarids	**May 1-8**
June Lyrids	June 10-21
Ophiuchids	June 17-26
Capricornids	July 15-August 15
Iota Aquarids	July 15-August 25
Perseids	**July 25-August 18**
Chi Cygnids	August 19-22
Orionids	October 16-26
Taurids	October 20 – November 30
Cepheids	November 7-11
Leonids	**November 14-19**
Geminids	**December 7-15**
Ursids	December 17-24

Note: Major showers printed in **bold**

K Index Values and Associated Effects on Radio Propagation

K index value	Geomagnetic Descriptor	Average occurrence per solar cycle	Typical effects on radio frequency communications
9	Extreme storm	4 days	HF propagation impossible, VHF aurora propagation
8	Severe	60 days	Sporadic HF propagation, VHF aurora propagation
7	Strong	130 days	Sporadic HF communications, VHF aurora propagation
6	Moderate	360 days	HF propagation fades at high frequency, some VHF aurora
5	Minor storm	900 days	short periods of VHF aurora propagation possible
4	Active	>1000 days	Generally no effect on HF propagation
3	Active	> 10,000 days	Generally no effect on HF propagation
2	Unsettled	> 10,000 days	Generally no effect on HF propagation
1	Quiet	> 10,000 days	Generally no effect on HF propagation
0	Quiet	> 10,000 days	Generally no effect on HF propagation

INDEX